1

Publishing House of the Slovak Academy of Sciences

Vydavateľstvo Slovenskej akadémie vied

Bratislava 1971

Nuclear Analytical Chemistry I

Scientific Editor

Prof. RNDr. R. Jirkovský

Reviewers

M. Křivánek, Ph.D.
Associate Prof. Ing. M. Rakovič, Ph.D.

English Translation

P. Tkáč

English Translation Edited by
Prof. RNDr. E. Körös, DSc. and Lucie Whittle

Nuclear Analytical Chemistry I

Introduction to Nuclear Analytical Chemistry

J. Tölgyessy
Š. Varga
and V. Kriváň

1971
University Park Press
Baltimore • London • Tokyo

Publishing House of the Slovak Academy of Sciences
Bratislava

Nuclear Analytical Chemistry I

Introduction to Nuclear Analytical Chemistry

by

Prof. Ing. J. Tölgyessy, Ph.D., DSc.

Prof. Ing. Š. Varga, Ph.D.

Department of Radiochemistry and Radiation Chemistry, Slovak Technical University, Bratislava, Czechoslovakia

and

Assoc. Prof. Ing. V. Kriváň, Ph.D.

Max-Planck-Institut für Metallforschung, Schwäbisch Gmünd, Germany

Library of Congress Catalog Card No. 70-158204

International Standard Book No. 0-8391-0079-5

Printed in Czechoslovakia

Contents

1. Introduction

J. TÖLGYESSY

The discovery of natural radioactivity by Henri Becquerel in 1896 and subsequent discoveries by Pierre Curie and Marie Curie-Skłodowska, opened a new area in the natural sciences. These discoveries started an intensive development in atomic physics which led to unusually interesting and far-reaching concepts in the most diverse branches of science. Even a glance at the pages of current scientific journals reveals the steadily growing number of contributions that, in one form or other, deal with radioactive radiation and its practical applications. The same is also true to a large extent for chemical analysis.

Radionuclides find an extensive and multiple application in chemical analyses for detecting and determining elements and compounds. Viewed as an analytical aid, radioactivity possesses exceptional advantages over all other analytical properties used in other methods. These comprise, in particular:

- an extraordinary sensitivity,
- a wide-range selectivity in detection and determination,
- speed and simplicity,
- elimination of the chemical handling of the sample (nondestructive analysis).

The first references to nuclear chemical analysis are seen in the work by Hevesy and Paneth [1] who in 1913 made use of a natural lead isotope (RaD) as indicator to determine the solubility of lead sulfide in water.

Prior to the discovery of artificial radioactivity, the number of radionuclides suitable for analytical purposes was rather limited. In order to make it possible to employ the above advantages in the detection and determination of substances by measurement of their radioactivity with simultaneous determination of other elements, analytical methods have been worked out that utilize natural radioactive elements as reagents. Radioactive substances began to be employed on a larger scale in analytical methods only after the production of the necessary artificial radionuclides in nuclear reactors had been worked out. Today radioisotopes of most of the elements are available, and as the original reason for the restriction in the development of these methods has ceased to exist, analytical methods using radionuclides have proved their worth in a wide range of applications. Recent years have witnessed the appearance of a number of new analytical procedures based on these principles. Therefore, today we may speak of a new scientific branch, viz., that of *nuclear analytical chemistry*.

Nuclear analytical chemistry forms part of applied nuclear chemistry, or of that section of analytical chemistry which in the qualitative or quantitative analysis of substances utilizes the nuclear characteristics of the corresponding nuclides. In the majority of cases, radioactive nuclides or nuclear radiation are involved, although stable nuclides may likewise be utilized (primarily as indicators) for analytical purposes. The difference lies solely in the methodology and the technique of the final measurement, i.e., the isotopic composition of the fractions is determined. In the case of light elements (hydrogen), this can be done by determining the density, and for the other elements by means of refractometric, interferometric or spectral methods, or, generally, by mass spectrography. The technique in the latter case is more exact and rather complex. This monograph series, *Nuclear Analytical Chemistry* is concerned before all else with the use of radioactive nuclides and nuclear radiation for analytical purposes.

An analytical utilization of radionuclides plays a considerable role in their overall application in chemistry. Nuclear analytical methods are gradually coming to be used as widely as any other physicochemical method. The importance of nuclear analytical procedures is proved also by the numerous scientific conferences and symposia devoted to this topic (organized by the International Atomic Energy Agency, e.g. the Symposium on the Use of Radionuclides in Science and Technology at The Hague in 1961; the Symposium on Radiochemical Analytical Methods, Salzburg, 1964, and others) or to selected methods of nuclear analytical chemistry, for instance, activation analysis (symposia on Modern Trends in Activation Analysis held in Vienna, at College Station, Texas, in Glasgow, etc.). Scientific meetings dealing with physicochemical analytical methods also include papers on nuclear analytical methods (for instance, at the IUPAC Conference, at national or international conferences).

At the Salzburg Symposium, Meinke [2] in his introductory speech aptly remarked that most of the nuclear analytical methods "have reached the stage of maturity during the last few years." Analytical methods that had formerly been judged unique, have now been refined on the technical and methodological side to such a degree that they are considered routine procedures in analytical chemistry and are applied even in industrial processes. Some of the nuclear analytical methods have already acquired their own tradition. Thus, for instance, activation analysis emerged from Prof. Hevesy's laboratory as a novelty 30 years ago. This particular analytical method has had a relatively lengthy inductive period in its development when compared with, say, polarography or nuclear magnetic resonance.

In spite of the well-documented and generally recognized importance of these methods, a dearth is still felt on a world-wide scale of books that would provide chemists and others interested in the subject with more systematic and deeper information on these relatively new but significant methods. Publications listed under Ref. [3–8] are exclusively concerned with this topic. Nevertheless, they do not deal with it to the required extent or depth.

The present monograph series *Nuclear Analytical Chemistry* is concerned with the use of radionuclides (eventually also of stable nuclides) and nuclear radiation in chemical analysis and certain related branches.

The main purpose of this series is to acquaint the reader with the principle underlying each individual method employed in nuclear analytical chemistry. In reality, these principles are not unduly complicated in themselves and by and large have been thoroughly investigated. However, when applied in practice they may give rise to various problems, mostly of a novel character, because of the peculiar nature of nuclear radiation.

Before proceeding further, we wish to emphasize the fact that this monograph series has been designed to embrace the subject matter in general, while providing at the same time deeper information on individual analytical methods. In processing the enormous amount of accessible material for this book we had in view two types of readers: the analytical chemist involved in practical work who wants to learn more about novel methods in analytical and nuclear chemistry from the point of view of his own professional activity; he will find the monographs handy reference books and, in fact, the numerous references cited in the text and the relatively exhaustive Subject Index are meant to familiarize him with the subject that is of interest to him, even though he does not intend to make a systematic reading of the book. The second type of reader is represented by students intending to major in chemistry at a university or a technical college; in their case, the monographs will provide a basis for a thorough grasp of the principles of nuclear analytical chemistry. Needless to say, a study of nuclear analytical chemistry presupposes a considerable standard of knowledge in "ordinary" chemistry.

In sorting out the subject matter of nuclear analytical chemistry we have decided to process the accessible material in the form of independent volumes, while keeping in mind the demand imposed by a purposeful and logical connection between individual volumes and chapters.

The first volume of this series, *Introduction to Nuclear Analytical Chemistry*, deals with the basic principles of nuclear physics and nuclear chemistry, detection and recording of nuclear radiation, of the peculiarities of work with radioactive elements, and the principles of indicator analytical methods.

The first chapter is devoted to the most important elementary concepts whose complete treatment belongs of course, to nuclear physics and nuclear chemistry. Although much of what is said in this first volume, and particularly in its first chapters, can be found in various textbooks, we nevertheless consider its inclusion in the present series as justified for two main reasons: first, we wish our monograph to form a complete entity, as far as this is possible within the framework of its principal purpose; second, many readers, especially chemists might feel disconcerted if they had to browse through standard textbooks on nuclear physics and nuclear chemistry in order to find the necessary basic information.

The second volume, *Radioactive Indicators in Chemical Analysis*, takes up practically all the methods in which radioactive nuclides (or even stable nuclides) are used as indicators (the so-called indicator methods). These are methods in which *the samples analyzed contain some natural radioactive nuclide* and thus become "naturally indicated," so that this radionuclide may be utilized for detection and determination of the corresponding element. Of great significance are *methods of isotope dilution analysis*, proposed by the late Nobel Prize winner, Prof. G. Hevesy, which permit accurate results to be obtained even when the component being determined cannot be isolated quantitatively. These methods are utilized in the analysis of inorganic substances, but their principal application is in the field of organic chemistry and biochemistry. In recent years the introduction of the substoichiometric principle to isotope dilution analysis proved to be very fruitful. A significant place in the group of radioindicator analytical methods is occupied by *radioreagent methods*, which comprise *simple radioreagent methods, methods of concentration dependent distribution*, and *radiorelease methods*. The concept of the *method of concentration dependent distribution* in quantitative analysis, using radionuclides, has been mentioned in the literature recently (M. Kyrš, 1966), even though determinations on this principle had been carried out more than ten years ago. Methods showing an unusually intensive development at the present moment are those of *radiometric titration*, involving the use of radioactive nuclides for end-point indication. A separate section of the volume is devoted to *radiochromatographic* and *radioelectrophoretic methods* (column adsorption-, ion exchange paper-, thin-layer-, gas-, radiochromatography, photoradiochromatography and electrophoresis of radioactive substances), although these procedures are used for the separation of components within the framework of the other radioindicator methods. Some of the reasons for including this separate and relatively extensive chapter lie in the peculiarities of the processing technique, above all, in the detection and quantitative evaluation of radiochromatograms and in the highly sophisticated instrumentation technique. Radioactive indicators are successfully employed *to check the accuracy of various analytical procedures*, to detect their deficiencies and thereby indirectly help to perfect the methods themselves. Of great significance also is the determination of important analytical constants (for instance solubility products, distribution coefficients, etc.). These aspects are dealt with in the last section of the second volume.

The third volume, *Radiochemical and Activation Analysis* deals with two important areas of nuclear analytical chemistry, namely methods of radiochemical analysis and activation analysis. The section dealing with *radiochemical analysis* follows the occurrence of radionuclides in life environment, preparation of samples for radiochemical analysis from natural sources (air, water, soil, vegetable and animal substances), methods and examples of analysis. Methods of radiochemical analysis are of great significance for the determination of radioactive substances in samples of the most diverse origin.

The most widely used nuclear analytical method is that of *activation analysis.* Its significance lies, above all, in its suitability for solving the question of trace analysis. This method is superior to all the analytical methods hitherto known, particularly because of its high sensitivity and specificity. The present-day standard in the preparation of the sources of activating radiation assures the appropriate conditions necessary for a wide-scale application of activation analytical procedures. The section of the volume dealing with activation analysis presents a complete survey on this important nuclear analytical method. It includes the theoretical basis of the method, its experimental technique which elucidates the entire working process from sampling and standard preparation, through activation, chemical sample processing, nondestructive analysis, right up to data processing and final evalution. The closing section points to the possible application of activation analysis in various areas of science and technology.

The fourth volume, *Analysis Based on the Interaction of Nuclear Radiation with Matter* deals with methods based on utilization of nuclear radiation when the sample contains no radionuclide at all. These involve the interaction processes of nuclear and X-ray radiations with the substance, leading to their absorption and backscattering. Radiation emitted by radionuclides and neutron radiation are employed most frequently. All these methods permit the analysis to be carried out without destruction of the substance examined. Here belong methods based on:

- the interaction between alpha radiation and the substance,
- absorption and backscattering of beta and gamma radiation,
- absorption of X-rays emitted by radionuclides,
- interaction between neutron radiation and substance.

The progress of the majority of these methods has lately become conspicuous by a decline, from a world-wide aspect, in the number of publications dealing with them. Their basic questions have already been processed in their essential features to such a depth that their practical application meets with no obstacle. In recent years a large number of methods have been developed in the field *of radionuclide fluorescence analyses.* The principle of these methods was known earlier, since they involve the use of photon radiation in the wave-length range of X-rays. The X-ray tubes can be conveniently replaced by X-ray or gamma sources. The fourth volume also includes new nuclear analytical methods, in particular, *analytical utilization of Mössbauer's effect*. This method is expected to extend rapidly, especially for the analysis of inorganic solids.

The fifth volume, *Tables, Nomograms and Schemas*, summarizes all the basic nuclear-chemical and nuclear-analytical data (e.g., nuclear characteristics of nuclides important for nuclear analytical chemistry, absorption and scattering, values for nuclear radiation in various substances, dosimetric tables and schemas, tables for radionuclide activation and decay, gamma spectra, ion exchange and their properties, solvent extraction, radiochemical separation of individual radionuclides, etc.) which the reader may find useful in setting up and carrying out individual analytical methods. These data are scattered throughout the literature and are often difficult to obtain.

The task of writing a comprehensive and systematic work on a branch of science that is still in a stage of development, has proved to be a highly exciting one indeed. The difficulty of preparing this series arises from the complex nature of a discipline in its full expansion, one whose exact delineation is still argued about. Hence, we shall be grateful for any remarks and suggestions that will enable us to improve this work.

Bratislava, 1967

References

1. Hevesy G., Paneth F., Z. anorg. Chem. 1913, **82,** 323.
2. Meinke W. W., in *Radiochemical Methods of Analysis*. IAEA, Vienna 1965, **1,** 13.
3. Broda E., Schönfeld T., *Radiochemische Methoden der Mikrochemie (Handbuch der mikrochemischen Methoden)*. Springer Verlag, Wien 1955.
4. Mikheeva L., Mikheev N. B., *Radioaktive Isotope in der analytischen Chemie*. Akademie Verlag, Berlin 1962.
5. Moses A. F., *Nuclear Techniques in Analytical Chemistry*. Pergamon Press, Oxford 1964.
6. Lambie D. A., *Techniques for the Use of Radioisotopes in Analysis*. E. and F. N. Spon, Ltd., London 1964.
7. Tölgyessy J., *Jadrové žiarenie v chemickej analýze (Nuclear Radiation in Chemical Analysis)*. SVTL, Bratislava, SNTL, Praha 1961.
8. Tölgyessy J., *Magsugárzás a kémiai analízisben (Nuclear Radiation in Chemical Analysis)*. Müszaki, Könyvkiadó, Budapest 1965.

2. Atomic Nucleus and Nuclear Radiation

J. TÖLGYESSY

2.1 Introduction

Atomic science is an area of physics that deals with the study of atoms and atomic complexes and research into their structure their properties and the processes involved. Just as two essential parts may be differentiated in the atom, so atomic science may be divided into two narrower areas: physics of the electronic shell (electron physics), and physics of the atomic nucleus (nuclear physics).

Nuclear physics, which deals with the structural composition and properties of atomic nuclei, holds an important place in modern science. It represents one of the youngest branches of physics, having begun its evolution as a science only toward the end of the 19th century. The history of atomic nucleus research forms one of the most fascinating chapters in the entire history of science. Even though a detailed treatment of the subject falls beyond the scope of this work, a concise review of the most important data from the history of nuclear physics cannot be omitted (Table 2.1) [1].

The purpose of this chapter is to present those principles and statements of elementary nuclear physics which are necessary to understand the principles underlying the use of radionuclides or nuclear radiation in nuclear analytical chemistry. Attention is here centered primarily on the structure of the atom, the laws governing radioactive decay, particular types of nuclear radiation, their interaction with the substance, and nuclear reactions. These concepts are important both from the viewpoint of the preparation of radioactive nuclides and the individual types of radiation, and from that of their detection and practical application in nuclear analytical chemistry.

2.2 Structure of the Atom

The concept of the composition of matter and the structure of the atom has passed through a lengthy evolution. At the beginning of this century the atom was still considered the smallest undivisible part of matter. The discovery of natural radioactivity and of the artificial transformation of elements has brought about deep changes in the earlier theory of the structure of the atom.

According to the nuclear model universally accepted today (E. Rutherford, 1912), two parts can be distinguished in the atom: a central (inner) positively charged part, represented by the *nucleus*, and an outer negatively charged one, formed by the *electronic shell.*

With the exception of the insignificant inner part—in terms of space—the electronic shell takes up practically the whole volume of the atom. The diameter of the central part is of the order of 10^{-4} of that of the whole atom (the central part represents about 10^{-12} of the total volume of the atom). However, although the electronic shell occupies nearly the whole volume of the atom, it contains only a negligible fraction of the entire matter of the atom (of the order between 10^{-3} and 10^{-4}). The electronic shell is the bearer of a negative electric charge and this is invariably an integral multiple of the elementary charge. The negative charge of the electronic shell counterbalances the effect of an equal-sized charge of the nucleus, so that in comparison with its own dimension (which is of the order of 10^{-10} m) it is electrically neutral. The number of electrons in the outer shell of the atom (in concordance with the electrical neutrality of the atom) in its normal state is given (similarly as the number of quanta of the electrical charge in the nucleus) by the atomic number Z.

The chemical properties of atoms are related to the electrons on the outer electronic shell. Periodicity of the chemical properties during a steady increase of the atomic number then testifies to the inner periodicity in the structure of the electronic shell, to the existence of the so-called shells, containing a certain number of electrons. The shell nearest to the nucleus is designated by K, the next by L and successively by M, N, ... Electrons of the same shell possess a relatively equal energy. When a chemical bond is formed, the outer electronic shell of the atom undergoes changes with far-reaching transformations in the properties relating to the outer electrons (optical spectra, conductivity, color, etc.). On the other hand, properties conditioned by the nucleus and the inner electrons (atomic weight, X-ray spectrum) are not altered by chemical processes.

The atomic nucleus constitutes the central core of the atom, and measures 10^{-14} m. Since the size of the entire atom is of the order of 10^{-10} m, the nucleus evidently represents only a negligible part of the whole atom. The nucleus is a quantum-mechanical system of nuclear particles, the so-called *nucleons*. The basic forms of

Table 2.1

Chronological Review of the Fundamental Discoveries in Nuclear Physics

1896	Becquerel	Discovery of radioactivity
1898	P. and M. Curie	Discovery of radium
1911	Wilson	Construction of the cloud chamber
1911	Rutherford	Discovery of the atomic nucleus
1913	Bohr	Quantum model of the atom
1915	Sommerfeld	Improved quantum model of the atom
1919	Rutherford	First artificial nuclear transmutation
1928	Geiger	Invention of Geiger-Müller counter
1930	Cockroft, Walton	Construction of the first particle accelerator
1930	Cockroft, Walton	First transmutation with artificially accelerated particles
1932	Lawrence	Construction of cyclotron
1932	Chadwick	Discovery of the neutron
1932	Anderson	Discovery of the positron
1932	Ivanenko, Heisenberg	Hypothesis on the composite nature of the nucleus
1934	F. and I. Joliot-Curie	Discovery of artificial radioactivity
1935	Yukawa	The meson theory of nuclear forces
1939	Bethe	Theory on thermonuclear reactions
1939	Hahn, Strassmann	Uranium fission by neutrons
1941	Kerst	Construction of betatron
1942	Fermi	Realization of chain reaction
1945	Veksler, McMillan	Discovery of the synchron accelerator principle
1945	Team of American scientists	Construction of the atomic bomb
1953	Teams of American and Soviet scientists	Construction of the hydrogen bomb
1954	Team of Soviet scientists	The first industrial nuclear power station set to work
1955	Chamberlain, Segré, Wiegand, Ypsilantis	Discovery of the antiproton
1956	Cork, Lambertson, Piccioni, Wenzel	Discovery of the antineutron

nucleons are positively charged *protons* and electrically neutral *neutrons*. The real mass of an atomic nucleus is smaller than the sum of the ground masses of its nucleons. The difference in mass is proportional to the *binding energy of the nucleus*. This is the energy that would be released (theoretically) during the synthesis of a nucleus, from nucleons. The bond energy of the nucleus, or the proportional diminution of its mass expresses the stability of the nucleus. The forces which bind nucleons together within the nucleus are called *inner nuclear forces*. The nature of these forces is specific, for classical physics does not know of any attraction existing between identical charges, or between charged and neutral particles. Even though the nature of these forces has so far not been definitely elucidated, we may presume that an essential role in the creation of bonds between the nucleons within the nuclei is played by a mutual interchange of π-mesons. It is further presumed that a nucleon consists of a central particle 10^{-16} m in size, surrounded by a "cloud" of mesons (the meson field).

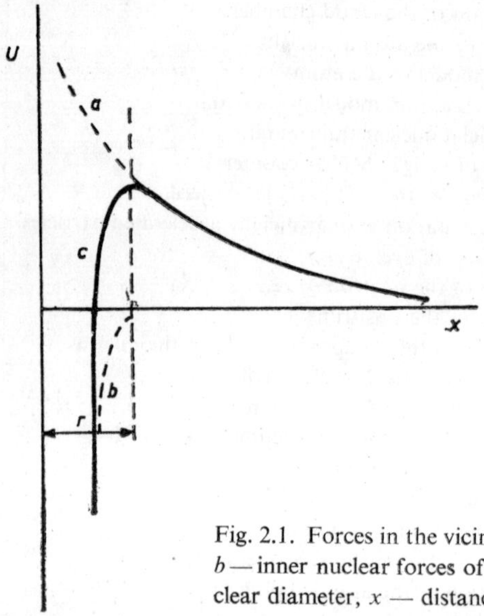

Fig. 2.1. Forces in the vicinity of a nucleus (a — Coulomb forces of repulsion, b — inner nuclear forces of attraction, c — resultant of both the forces, r — nuclear diameter, x — distance from the center of the nucleus, U — potential).

Our conception of the spatial arrangement of an atomic nucleus is that it has a spherical shape as its basis. The conditions prevailing in the forces of the nucleus are illustrated in Fig. 2.1. Forces for positively charged particles in the neighborhood of the nucleus are made up of two components: first, the Coulomb repelling forces (curve a) which decrease with the square of the distance; second, the inner-nuclear attracting forces (curve b) which are extraordinarily great in the vicinity of the nucleus but decrease considerably with increasing distance. The resultant of both these forces (curve c) represents the potential barrier around the atomic nucleus. The effective radius of the nucleus (r) corresponds to the distance of the maximum potential barrier from the center of the nucleus.

Atomic nuclei may be characterized by means of the following numbers:

Atomic number (Z) indicates the number of protons in the nucleus and thereby also the number of the elementary charges of the nucleus. The same number of electrons is present in the electronic shell.

Mass (nucleonic) number (A) indicates the sum of protons and neutrons in the nucleus ($A = Z + N$).

Neutron number (N) indicates the number of neutrons in the nucleus ($N = A - Z$).

Isotopic number (I) indicates the difference between the number of neutrons and protons ($I = N - Z$). It gives the excess of neutrons in the nucleus with respect to the number of protons.

Atomic nuclei may show agreement in some of the numbers and this helps to differentiate several types of atomic nuclei (nuclides). Nuclei having the same A values but differing in N and Z are termed *isobars*; those with equal Z and differing in N are known as *isotopes*. *Nuclear isomers* (isobaric isotopes) are atomic nuclei having identical Z and A and differing solely by the energetic states of their nuclei (parameters of nuclei transformations).

Let us now turn to the *inner characteristics of a nucleus*. When two or more nucleons are mutually bound in a nucleus, then every quantum state of such a bond system is called the *energy level of the nucleus*. The existence of energy levels in nuclei has been proved by various experiments. These levels cannot be ascribed to a single nucleon, but belong to the nucleus as a whole. Consequently the energy spectrum of a nucleus in no way resembles an atomic spectrum, rather it is reminiscent of the spectrum of a substance in a condensed state, for instance that of a crystalline solid.

The nucleon system can absorb or emit energy. In view of the quantum-mechanical character of this system, the amounts of energy (emitted or absorbed) are quantified (discretely). A result of energy absorption is a relatively "long-lived" state of the nucleus. These so-called *excited states*, also termed excited energy levels of the nucleus, correspond to various groups of nucleons in the nucleus.

Energy levels are often represented graphically: Values of the corresponding energy level are plotted on the vertical axis of the diagram (energy for the basic state is taken as 0). This gives some sort of a diagram of energy levels (Fig. 2.2). In the early stages of investigation into the nucleus, its various energy levels were studied only during the process of nuclear decay. Today systematic information is supplied by nuclear reactions. Transitions between pairs of nuclear energy levels give rise to radiations, which are usually measurable [2].

Energy levels of atomic nuclei, in the case of low energy of the nucleus, are distributed over great distances, but with an increasing inner energy of the nucleus their density grows rapidly. For high energy levels of the order of 15–20 MeV and above, energy levels are so close to each other that their spectrum may be considered as continuous.

The decreasing average distance between the levels and their increasing number with an increased excitation energy of the nucleus is shown in Fig. 2.3. To give a more graphic view, the levels of strongly excited nuclei are shown "under a magnifying glass." The widths of the levels (see p. 71) expand with increasing excitation energy, so that the strongly excited nucleus is far more short-lived than one at a lower energy level. When a heavy nucleus is excited to an energy approaching 10 MeV, the widths of the levels are, for the most part, greater than their mutual distances, so that various levels overlap [3].

Every inner state of the nucleus (arranged grouping of bound nucleons) may be described by a series of quantum parameters, for instance, by excitation energy, lifetime (the so-called width of the level), nuclear spin, and others.

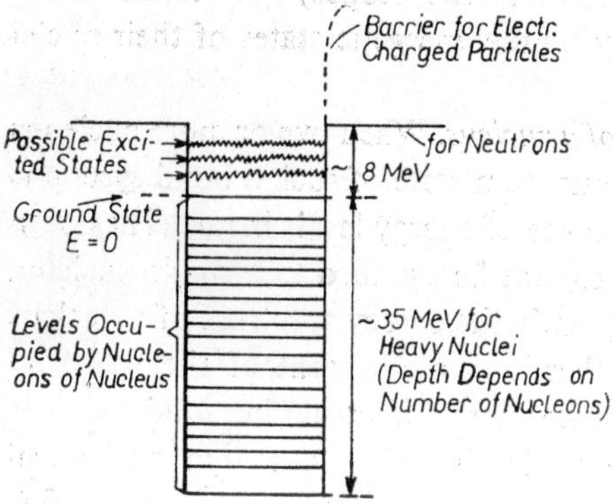

Fig. 2.2. Potential well and energy levels of the nucleus.

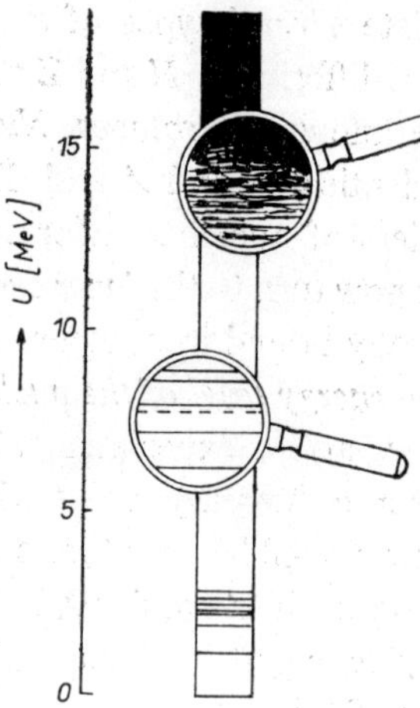

Fig. 2.3. Dependence of the distance between levels on the excitation energy of the nucleus.

2.3 Radioactivity

So far about 1300 atomic nuclei are known, differing in their charge, masses or both. Approximately 300 of these nuclei are *stable*. Radioactive nuclei decay spontaneously (i.e. from inherent causes) and during the process of emitting the so-called radioactive radiation, change into other nuclei.

Radioactive decay is not affected either by temperature, pressure, or any other external factor. Radioactive radiation originates from the atomic nucleus. Whether the atomic nucleus is stable or not, that is, whether it is subject to radioactive decay or not, how it decays and at what rate, all this depends upon its composition and state.

An atomic nucleus is more or less stable depending on the ratio of the number of protons to that of neutrons. In the Fig. 2.4 the abscissa carries the number of protons Z, and the ordinate that of neutrons ($N = A - Z$). Full circles in the diagram designate stable isotopes and empty circles radioactive ones. Stable isotopes of the given nucleus are on the vertical lines (constant Z), stable isotopes on the horizontal lines (constant N) and stable isobars are shown by slanting lines. The range of stability as shown in the figure, is relatively narrow. For a low mass number $N/Z = 1$, and in the case of high mass numbers this ratio increases approximately to 1.6. The circles to the right of stable isotopes denote unstable nuclei with excess of protons,

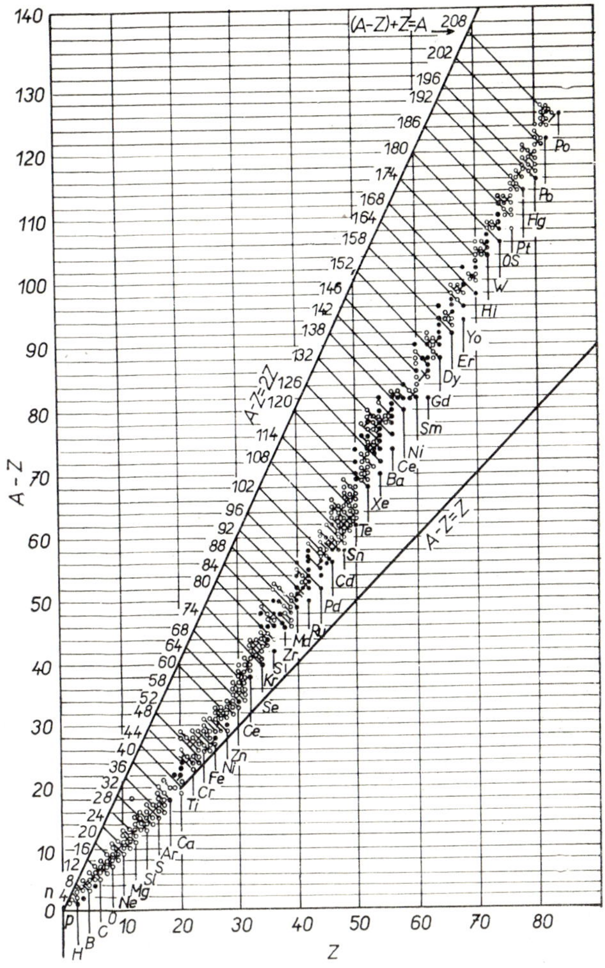

Fig. 2.4. Stable and radioactive nuclides.

those to their left, nuclei with excess of neutrons. In both these cases such nuclei strive by means of some type of radioactive decay to equalize these excesses and thus attain the region of stability.

Figure 2.5 presents a more detailed illustration of the first part of the diagram. It shows data on the occurrence of stable isotopes (percentage content of a certain stable isotope of the given element)—on which may be a measure of the stability of

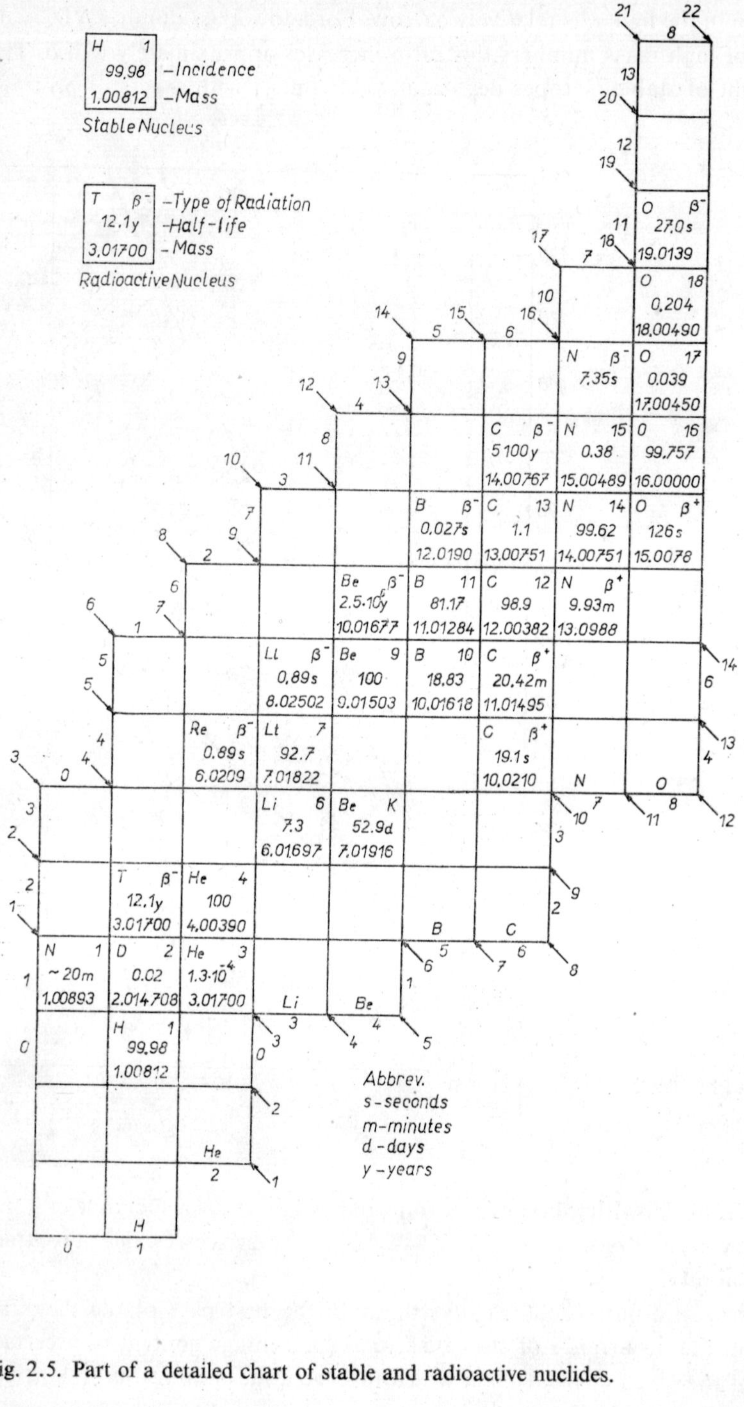

Fig. 2.5. Part of a detailed chart of stable and radioactive nuclides.

the nucleus – on the type of radioactivity of unstable isotopes and their half-life. Nuclei with an excess of neutrons emit excess negative electrons (β^- radioactivity), nuclei with an excess of protons excess emit positive electrons (β^+ radioactivity).

Five basic types of radioactive transformations are known at present. In four of them, that is alpha decay (emission of alpha particles), β^- and β^+ decay (emission of negatrons and positrons from the nucleus), electron capture (seizure of electrons from the shell nearest to the nucleus) and spontaneous fission, a change of nuclear charge occurs and the transformation of one chemical element to another takes place. The fifth type of radioactive transformation, that is emission of gamma quanta, is related only to the transition of nuclei from one level to another and is not accompanied by a change in the number of protons or neutrons in the nucleus.

From the viewpoint of the chemical properties of an element the decisive factor resides essentially in whether its atoms are radioactive or not. As long as a radioactive isotope fails to decay, it preserves its properties just a stable one does.

2.3.1 The Law of Radioactive Decay

Spontaneous transformation of a radioactive element leads to its decrease. If the sample of a substance contains a considerable amount of radioactive atoms of a certain element, then every nucleus of the atoms present has a certain probability that in a given time interval its transmutation will occur. The radioactive decay has a statistical character. In view of the fact that a great number of radioactive atoms is always involved, their decay can be described with a high degree of precision.

The correlation between the decrease in the original amount of radioactive atoms and the time of decay is given by the law of decay

$$\mathrm{d}N = -\lambda N \,.\, \mathrm{d}t, \tag{2.1}$$

where N is the number of radioactive atoms in time t, $\mathrm{d}N$ the number of atoms which will decay during time $\mathrm{d}t$, and λ the proportionality constant (decay constant) – characterizes the rate of nuclear decay.

By integrating equation (2.1) with the initial conditions $t = 0$, $N = N_0$ we obtain

$$N_t = N_0 \,.\, \mathrm{e}^{-\lambda t}, \tag{2.2}$$

as the number of radioactive atoms N_t at any given time t, to be. This is the time law of radioactive transformation according to which the decrease of the number of radioactive atoms in time obeys an exponential law (Fig. 2.6). Since the activity of the radioactive substance is directly proportional to the number of atoms, equation (2.2) can be written as follows:

$$I_t = I_0 \, . \, e^{-\lambda t}, \tag{2.3}$$

where I_0 and I_t are activities at the corresponding time, expressed by the number of decay in a unit time, or by the intensity of the ionization current (p. 84).

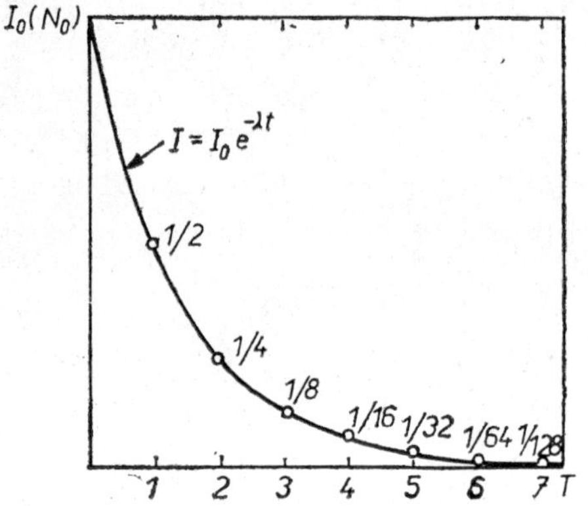

Fig. 2.6. Decrease in radiation intensity I (or the number of radioactive atoms N) in time.

The rate of radioactive transformation is often characterized by means of *half-life T*, defined as the time in which half the amount of radioactive atoms decay. Hence, for $t = T$, $N = N_0/2$. By substituting these values in equation (2.2), we obtain

$$T = \frac{0.693}{\lambda}. \tag{2.4}$$

The half-life may conveniently be calculated from the kinetic curve of radioactive transformation (Fig. 2.6).

A given radionuclide may also be characterized, besides half-life, by *the average life* (τ) of its atoms, which is given by the reciprocal value of the decay constant

$$\tau = \frac{1}{\lambda} = 1.4428 \; T. \tag{2.5}$$

The average life of a radioactive substance is the time during which the substance would completely decay if the number of atoms decaying during the first second, remained constantly the same. For instance, the half-life of radium T is 1590 years, the average life of its atoms (τ) is 2290 years. During the average life, the original number of atoms diminished by 37 %.

The values λ, T and τ are characteristic for every radioactive substance and they cannot be altered by any external factor whatsoever.

2.3.2 Radioactive Equilibrium and Decay Series

In the majority of cases the disintegration product of a radioactive nuclide (the so-called daughter nuclide) is as radioactive as its parent nuclide. In such a case the relationship for radioactive decay is somewhat more complex and must be expressed as

$$\frac{dN_2}{dt} = \lambda_1 N_1 - \lambda_2 N_2, \tag{2.6}$$

where subscripts 1 and 2 refer to the parent and daughter element respectively.

It ensues from relationship (2.6) that the daughter nuclide will accumulate until $\lambda_1 N_1$ is greater than $\lambda_2 N_2$. The magnitude dN_2/dt goes on diminishing all the time, for N_1 decreases and N_2 augments. The latter will augment until dN_2/dt equals zero.

The condition

$$\lambda_1 N_1 - \lambda_2 N_2 = 0 \tag{2.7}$$

is called *the condition of radioactive equilibrium*. During radioactive equilibrium the number of atoms decaying in the daughter product is equal to the number that is formed during the given time interval from the parent substance. During radioactive equilibrium the relation between the number of daughter atoms and parent nuclei is

$$\frac{N_2}{N_1} = \frac{\lambda_1}{\lambda_2} = \text{Constant}, \tag{2.8}$$

i.e. it is independent of time.

Relationship (2.8) may also be expressed in the form

$$\frac{N_2}{N_1} = \frac{T_2}{T_1}. \tag{2.9}$$

Hence, in a radioactive equilibrium, the proportion of atoms in the parent and the daughter element is given by the ratio of the corresponding half-lives.

If condition (2.6) for radioactive equilibrium is not fulfilled, i.e. $dN_2/dt \neq 0$, then, for long half-life parent elements the following equation is valid

$$N_2 = N_1^0 \frac{\lambda_1}{\lambda_2}(1 - e^{-\lambda_2 t}), \tag{2.10}$$

where N_1^0 is the value of N_1 when $t = 0$ (initial quantity). For a parent element of relatively short half-life ($\lambda_1 > \lambda_2$) the following relationship is valid

$$\frac{dN_2}{dt} = \lambda_1 N_1^0 . e^{-\lambda_1 t} - \lambda_2 N_2, \tag{2.11}$$

or

$$N_2 = N_1^0 \frac{T_2}{T_1 - T_2} (e^{-\lambda_1 t} - e^{-\lambda_2 t}). \tag{2.12}$$

If the half-life T_1 of a parent element is considerably greater than T_2 of the daughter element, then, for a sufficiently higher value t, the equation

$$N_2 = N_1^0 \frac{T_2}{T_1 - T_2} e^{-\lambda_1 t} \tag{2.13}$$

will hold.

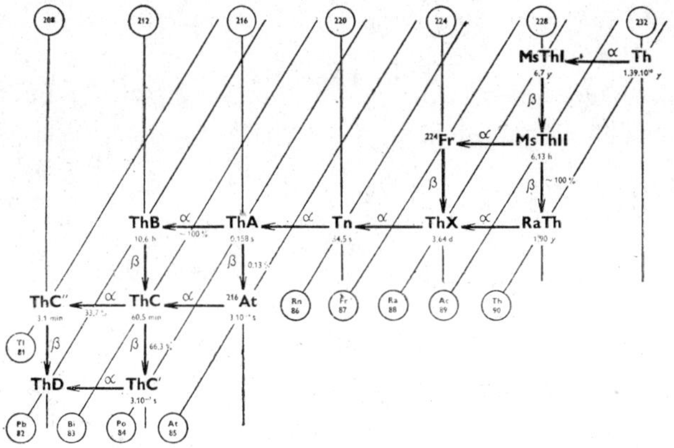

Fig. 2.7. Thorium disintegration series.

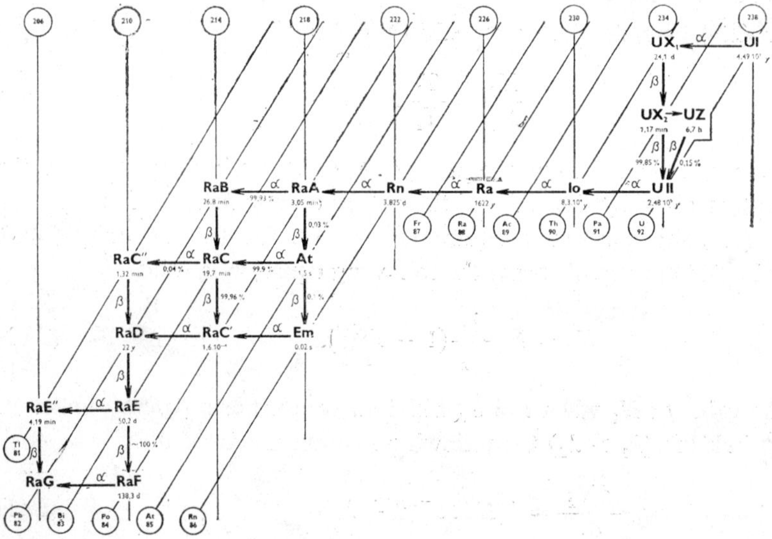

Fig. 2.8. Uranium disintegration series.

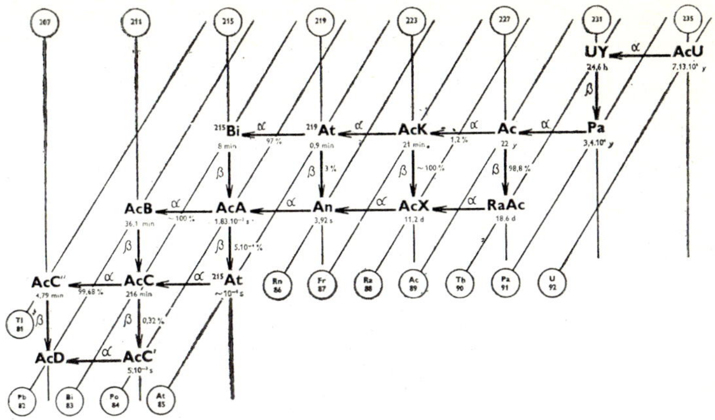

Fig. 2.9. Actinium disintegration series.

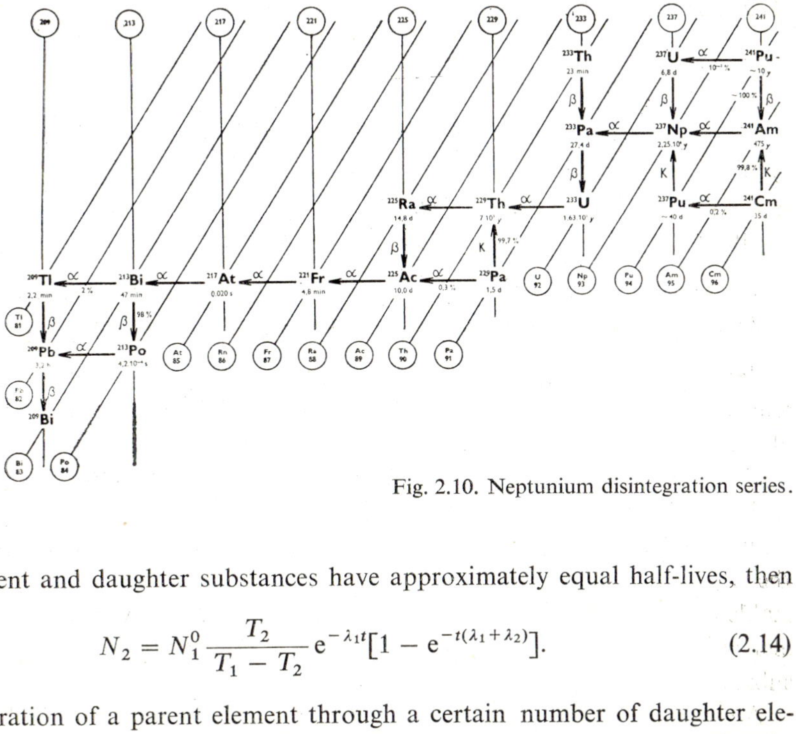

Fig. 2.10. Neptunium disintegration series.

If the parent and daughter substances have approximately equal half-lives, then

$$N_2 = N_1^0 \frac{T_2}{T_1 - T_2} \mathrm{e}^{-\lambda_1 t}[1 - \mathrm{e}^{-t(\lambda_1 + \lambda_2)}]. \tag{2.14}$$

The disintegration of a parent element through a certain number of daughter elements will lead to a stable nuclide. All the nuclides of this natural radioactive decomposition process belong to a *radioactive family* or *decay series.*

In every series the atomic number Z of the element changes because of individual alpha or beta disintegration according to the *displacement laws of Soddy and Fajans.* These displacements are illustrated in Figs. 2.7, 2.8, 2.9. Because by emitting an alpha or a beta particle the mass number of a nuclide changes either by 4 units or remains unaltered, individual members of every series (their A) can be expressed by the general formula $4n + a$, where n is an integer and a is either 0, 2 or 3:

1. Thorium series, $4n$, starts with thorium $^{232}_{90}Th$ and ends with thorium D ($^{208}_{82}Pb$).
2. Uranium series, $4n + 2$, starts with uranium I ($^{238}_{92}U$) and ends with radium G ($^{206}_{82}Pb$).
3. Actinium series, $4n + 3$, starts with actinouranium ($^{235}_{92}U$) and ends with actinium D ($^{207}_{82}Pb$).

Series $4n + 1$ was discovered much later than the artificial neptunium series (Fig. 2.10).

2.4 Nuclear Radiation

Practically every process of nuclear transmutation is related to the emission of particles or photons originated in the atomic nucleus and these represent a particular type of radiation. This radiation has its origin in the atomic nucleus and is therefore generally called *nuclear radiation.*

Nuclear radiation may take place in two ways:

1. in natural radioactivity by a spontaneous disintegration of atomic nuclei,
2. by the interaction of particles or photons with atomic nuclei.

Spontaneous disintegration of atomic nuclei occurs with certain elements in nature which change into other elements close to them in the periodic table, while emitting alpha, beta or gamma radiation. Nuclear radiation is accompanied by a series of nuclear processes which may be affected by bombarding atoms of a given element with particles; this sets up an interaction between them. These particles may come from natural radioactive substances, they may be artificially accelerated, or they may reach the earth's atmosphere from cosmic radiation. Certain interactions give rise to unstable nuclei which then disintegrate spontaneously. In this way, artificial radioactive elements can be produced with properties similar to those in natural radioactive substances.

A common feature of all the types of nuclear radiation is their high energy. This high energy content gives rise to a special form of action on a substance. When passing through matter, nuclear radiation induces, among other processes, that of ionization. Hence, nuclear radiation is often called also *ionizing radiation.* The ionization capacity of radiation may at the same time be a manifestation of its direct and indirect action on the substance.

2.4.1 Types of Nuclear Radiation

The basic types of nuclear radiation which are used the most frequently in nuclear analytical chemistry are the alpha, beta, gamma, neutron, deuteron and proton radiations.

2.4.1.1 Alpha Radiation

Alpha radiation is made up of positively charged particles (i.e., of bivalent positive helium ions, ${}_{2}^{4}He^{2+}$). In consequence of the emission of four nucleons (two protons and two neutrons) the mass number decreases by four units and the atomic number by two units:

$$ {}_{Z}^{A}X = {}_{Z-2}^{A-4}Y + {}_{2}^{4}He, \tag{2.15} $$

where X is the parent element and Y the daughter element.

The binding energy of the nucleons in the alpha particles is enormous (more than 28 MeV) and therefore the alpha particles in their mutual interaction with matter always behave as single particles. The alpha particles usually leave the nucleus of the atom in monoenergetic groups. For the most part their emission is accompanied by the emission of one or several quanta of gamma radiation. Figure 2.11 shows the disintegra-

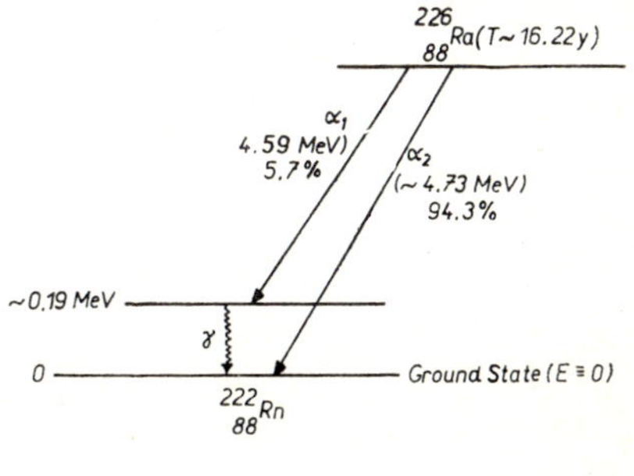

Fig. 2.11. Radium ${}^{226}Ra$ alpha decay.

tion schema for radium ${}^{226}Ra$ to radon ${}^{222}Rn$. The energy values indicate the energy levels of atomic nuclei. The energy of the alpha particles cannot be judged from their difference, for 94.3 % of the alpha decays of ${}_{88}^{226}Ra$ result in ${}_{86}^{222}Rn$; 5.7 % of the decays, however, result in excited ${}_{86}^{222}Rn$ nuclei which in their transition to a lower energy emit photons.

Alpha radiation is emitted by the majority of natural radioactive elements; it ensues from artificial radiation or is produced by an acceleration of helions in cyclotrons, synchrotrons, synchrophasotrons or similar accelerating installations.

Alpha particles have a characteristic initial velocity of the order of 10^7 m sec^{-1}, and energies between 4 and 9 MeV. The energy of alpha radiation depends on the nature of the radioactive element. The energies of alpha particles can be measured by determining the radius of their circular pathway in a magnetic field, or less accurately, by their R_{max} range.

2.4.1.2 Beta Radiation

Beta radioactive transformations are characterized by the emission of negative or positive electrons (negatron or positron) from the nucleus, or by the capture of a negatron by the nucleus.

Beta radiation of natural radioactive elements consists of negative electrons ($_{-1}^{0}e$, β^-), the speed of beta particles moving at a rate approaching the velocity of light. Certain radioactive elements, however, emit beta-radiation made up also of positrons ($_{+1}^{0}e$, β^+). Beta particles emitted by a source exhibit a continuous distribution of energy. The energies range from very small values up to a definite maximum E_{max}. Plotting the relative number of particles possessing a particular energy against the energy, a curve resembling a Gaussian approximation is obtained, the peak of the curve is shifted towards the low energy values. Figure 2.12 illustrates the beta spec-

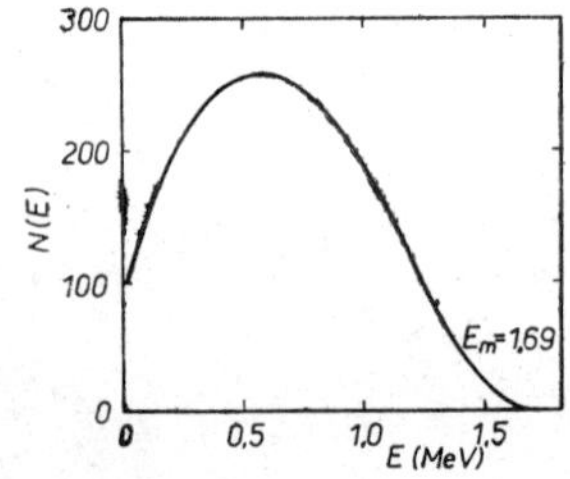

Fig. 2.12. Beta spectrum of ^{32}P [N(E) is the number of beta particles of energy E].

trum of ^{32}P. The maximum energy of beta radiation is easily measurable. To characterize a given radionuclide, the maximum energy of its beta radiation is always given along with some other nuclear data. The mean beta radiation energy is equal approximately to one-third of the maximum energy

$$\bar{E} \doteq \frac{E_{max}}{3}. \tag{2.16}$$

The maximum energy of beta radiation is inversely proportional to the half-life of the radioactive elements.

Beta particles are not a constituent of the atomic nucleus but are produced during the beta decay. This involves a mutual conversion of nucleons within the atomic nucleus: neutron conversion to proton, or vice versa. In all these transformations, neutrino ($^0_0\nu$) or antineutrino ($^0_0\bar{\nu}$) is emitted from the atomic nucleus. By agreement, neutrino designates the electrically neutral particle emitted duıing a positive beta decay and *K*-capture, while antineutrino is the particle emitted in a negative beta decay.

Neutrino is an electrically neutral particle having very small rest mass. Experimental pıoof for the existence of neutrino came as late as 1956. To account for the beta particle energy spectrum the formation of the neutrino was postulated. The sum of energies of beta particles (E_e) and of the corresponding neutrino (E_ν) is constant and is equal to the maximum energy of the continuous spectrum

$$E_{max} = E_e + E_\nu. \tag{2.17}$$

This is the cause of the continuous shape of the energy spectra of beta particles.

The negative (electron, negatron) beta disintegration is the most important and the most widespread form of beta decay. It takes place whenever the ratio of the number of neutrons and protons within the nucleus of a given nuclide exceeds a certain limit. In a negative beta decay one neutron in the atomic nucleus is converted to a proton, a negatively charged electron (negatron) and an antineutrino, according to the schema

$$^1_0n \rightarrow {}^1_1p + {}^{\ 0}_{-1}e + {}^0_0\bar{\nu}. \tag{2.18}$$

As a result of negative beta decay the atomic number increases by one while the mass number remains unchanged:

$$^A_ZX \rightarrow {}^{\ \ \ A}_{Z+1}Y + {}^{\ 0}_{-1}e. \tag{2.19}$$

For example:

$$^{32}_{15}P \rightarrow {}^{32}_{16}S + {}^{\ 0}_{-1}e.$$

Neutrino is not included in the equations.

The disintegration schema for ^{32}P decay is shown in Fig. 2.13.

A negative beta decay is characteristic for both natural and artificial radioactive elements.

The positive (positron) beta disintegration is related to the emission of positrons. It is typical for artificial radioactive nuclides whose neutron–to–proton ratio is lower than that corresponding to the stable nuclides of a given element. In a positive beta decay one proton in the nucleus is converted to a neutron, positron and neutrino:

$$^1_1p \rightarrow {}^1_1n + {}^{\ 0}_{+1}e + {}^0_0\bar{\nu}. \tag{2.20}$$

From the above it ensues that the atomic number in this decay decreases by one unit and the mass number remains unchanged:

$${}^{A}_{Z}\text{X} \rightarrow {}^{A}_{Z-1}\text{Y} + {}^{0}_{+1}e. \qquad (2.21)$$

For example:

$${}^{22}_{11}\text{Na} \rightarrow {}^{22}_{10}\text{Ne} + {}^{0}_{+1}e.$$

The disintegration schema for ^{22}Na decay is shown in Fig. 2.14.

The positive beta decay is always accompanied by the emission of gamma quanta.

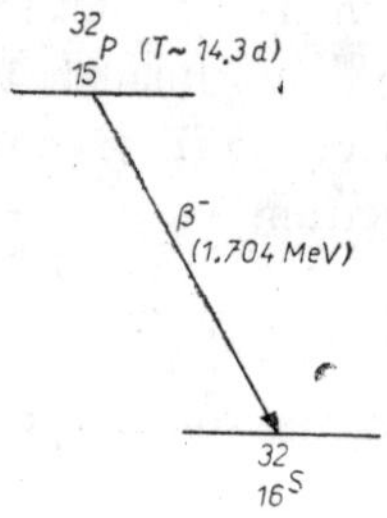

Fig. 2.13. Schema of radioactive decay for ^{32}P.

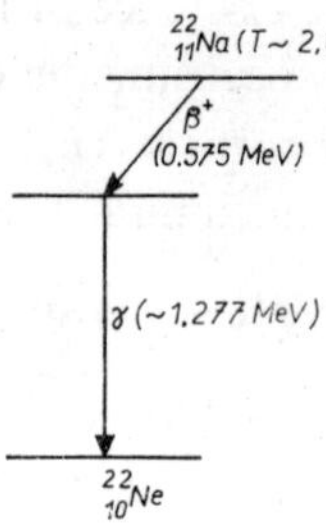

Fig. 2.14. Schema of radioactive decay for ^{22}Na.

A slowed-down positron at the end of its path easily reacts with a negatron, giving rise to two gamma quanta emitted in opposite directions. The energy of the quanta thus formed is equivalent to the rest mass of the original particles (0.51 MeV). The gamma radiation thus produced is termed *annihilation radiation.*

K-capture is a third variant of beta decay in which one electron from the electron shell (most frequently from the *K* shell) is captured by the nucleus and the proton is converted to a neutron:

$${}^{1}_{1}p + {}^{0}_{-1}e \rightarrow {}^{1}_{0}n + {}^{0}_{0}\bar{\nu} \qquad (2.22)$$

hence the atomic number decreases by one unit and the mass number remains unchanged, similar to that observed in a positron disintegration process:

$${}^{A}_{Z}\text{X} + {}^{0}_{-1}e \rightarrow {}^{A}_{Z-1}\text{Y}. \qquad (2.23)$$

For example:

$${}^{7}_{4}\text{Be} + {}^{0}_{-1}e \rightarrow {}^{7}_{3}\text{Li}.$$

K-capture is represented by the decay schema in Fig. 2.15. The removal of an extranuclear electron leaves a vacancy in the appropriate quantum level (*K*, *L*, ...). An electron from one of the higher energy levels will then immediately move in to fill the vacant position, and the excess energy will be emitted as the corresponding characteristic X-ray. This may eventually provoke a photoelectrical phenomenon in its own electron shell (Auger electrons). This process is schematically represented in Fig. 2.16.

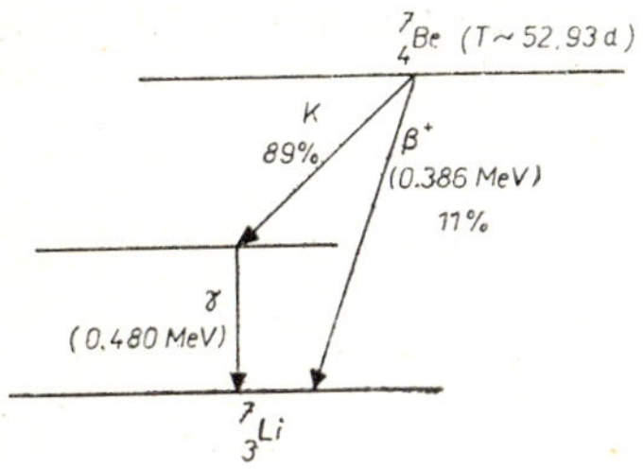

Fig. 2.15. Schema of radioactive decay for ^{7}Be.

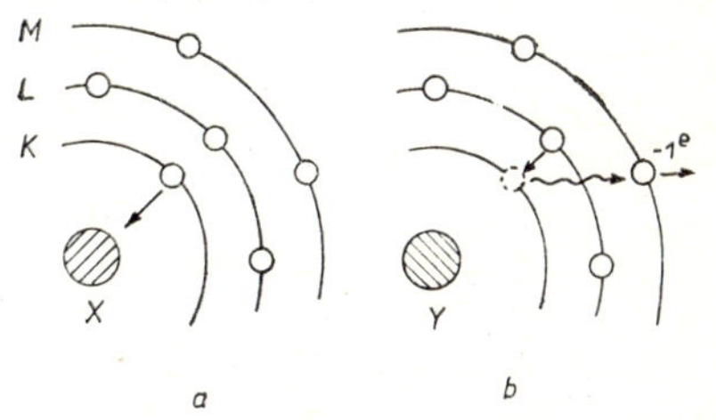

Fig. 2.16. Schematic model of the processes involved in an electron capture by the nucleus (*a*—absorption of planetary electron *K*, *b* — electron transition from another path, with emission of the characteristic X-radiation and ejection of Auger electrons from the atomic shell).

2.4.1.3 Gamma Radiation

Gamma radiations are highly penetrating electromagnetic oscillations similar to X-rays, but of shorter wavelength (0.005 to 0.4 Å). They are streams of photons with an energy between 0.05 and 5 MeV. For this reason gamma radiations do not affect either the mass or the atomic number of the parent radionuclide

$$^{A}_{Z}X \rightarrow \, ^{A}_{Z}X + \, ^{0}_{0}\gamma. \tag{2.24}$$

Like all electromagnetic waves, gamma rays also travel with the velocity of light and their energy is expressed by the known relation

$$E_\gamma = h \, . \, \nu_\gamma = h \frac{c}{\lambda_\gamma}, \tag{2.25}$$

where E_γ [erg] is the quantum energy, h is 6.62×10^{-27} erg sec^{-1} (Planck's constant), ν_γ [l sec^{-1}] the frequency of the gamma ray, and c the velocity of light in vacuo.

The correlation between wavelength λ_γ [Å] and energy E_γ [MeV] of gamma radiation is given by the following relation

$$\lambda_\gamma = \frac{1.238}{E_\gamma} 10^{-2}. \tag{2.26}$$

Gamma radiation may originate from natural or artificial radionuclides, or as secondary radiation from electron accelerators. The terms gamma and X-rays, respectively refer only to the origin of radiation. If the electromagnetic radiation originates within the nucleus, it is termed gamma radiation; if, however, it does so within the electronic shell, the term X-rays (Roentgen) is employed. Gamma radiations are as-

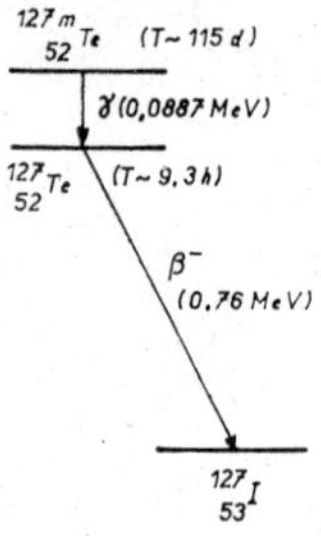

Fig. 2.17. Schema of radioactive decay with isomeric transition.

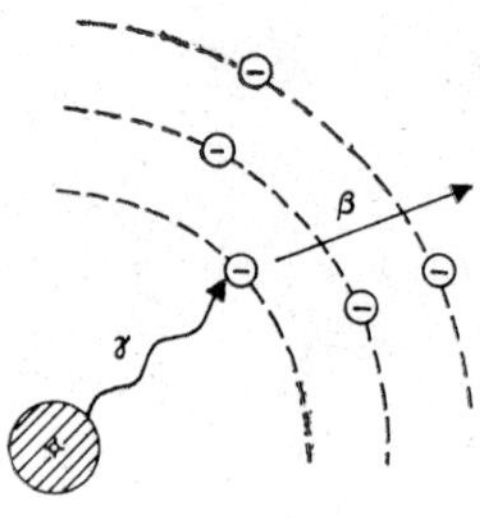

Fig. 2.18. Secondary beta radiation produced during emission of gamma quanta from atomic nucleus.

sociated with transitions between nuclear energy levels; the atomic nucleus passes from a higher energy into an isomeric state (a nucleus with the same atomic and mass number) in which it is either stable, or loses further energy in one of the ways described above, or in their combinations.

Figure 2.17 illustrates the disintegration schema of an isomeric transition.

Gamma radiations frequently accompany alpha or beta decays. A gamma photon emerging from a nucleus, in its passage through the electronic shell can transfer its energy in the K shell to a K electron without leaving the atom. The electron which interacts with the gamma photon is ejected from the atom as shown in Fig. 2.18. The energy of the gamma photon is consumed partly to overcome the binding energy of the electron on the atom and partly it appears as kinetic energy of the ejected electron. Simultaneously, however, one electron from the L or the M shell fills the place left vacant by the K electron, and a soft secondary gamma radiation is observed. This phenomenon is called *an internal conversion.* Internal conversion then manifests itself as a negative beta radiation accompanied by an X-radiation.

Photons emitted in nuclear processes are often termed differently according to the manner in which they are produced [2].

a) *Gamma radiation* is an electromagnetic radiation associated with transitions between nuclear energy levels.

b) "*Bremsstrahlung*" is an electromagnetic radiation appearing when a charged particle is either accelerated or slowed down. The energy of this radiation is not distinct for all the photons taking part in the "Bremsstrahlung" process but it is continuous.

c) *Characteristic X-rays* originate when inner orbital electrons return from an excited to the ground state.

d) *Annihilation radiation* appears when positive and negative charges (e.g., a negatron and a positron) neutralize each other and the particles are annihilated, leaving only energy in the form of electromagnetic radiation.

Electromagnetic radiation is involved in all the transition types, with different energy levels, however. The interaction of these photons with matter does not, in view of their common character, depend upon the manner in which this radiation is induced, but solely upon the energy of the photon.

2.4.1.4 Neutron Radiation

The natural source of neutrons is cosmic radiation. But neutrons appear also in artificial nuclear reactions, in laboratory neutron sources, in nuclear reactions produced by means of particles from accelerators or in nuclear reactors.

The neutron carries no electric charge. Its mass is 1.008 98 m.u.

The behavior of a neutron depends upon whether it is free, or bound in the nucleus. If a neutron is in the nucleus it may be stable, but in a free state it converts into a proton and an electron.

$$ {}^{1}_{0}n \rightarrow {}^{1}_{1}\mathrm{H} + {}^{0}_{-1}e. \tag{2.27} $$

The wavelength of neutron radiation λ_n is

$$ \lambda_n = \frac{h}{m_n \cdot v_n} = \frac{h}{\sqrt{2m_n \cdot E_n}}, \tag{2.28} $$

where h is Planck's constant, m_n the mass, v_n the velocity, and E_n the neutron energy.

Neutrons are diffracted by crystals in the same way as are X-rays and hence, neutron spectroscopy can be used for the determination of crystal structures.

According to their energy value neutrons can be classified as

very fast – $E > 10$ MeV,
fast – $E = 10$ to 0.5 MeV (at 1 MeV their velocity is 1.4×10^4 km sec^{-1}),
medium fast – $E = 500$ to 1 keV,
slow – $E < 1$ keV.

Slow neutrons are further subdivided into so-called *resonance* ($E = 1$ keV to 0.1 eV) and *thermal* neutrons ($E < 0.1$ eV; at 0.025 eV, which corresponds to the average kinetic energy of molecules at ordinary temperature whose velocity is about 2 km sec^{1-}).

The individual terms for the neutron groups have no strictly defined signification: a considerable lack of uniformity is apparent from the literature as to their usage and consequently only the context enables one to infer which neutrons are involved.

The relationship between thermal neutrons and their velocity and mass is given by the expression

$$E_t = \frac{1}{2} m_n \,.\, v_t^2 = \frac{3}{2} kT, \tag{2.29}$$

where E_t is the energy of thermal neutrons [eV], v_t the velocity of thermal neutrons [cm sec^{-1}], m_n the neutron mass [g], T the environmental temperature [°K], and k Boltzmann's constant [(1.380 44 ± 0.000 07) × 10^{-23} J grad^{-1}].

2.4.1.5 Proton Radiation

Protons are, in their essence, the atomic nuclei of the lightest hydrogen isotope ($^1_1$H). E. Marsden (1914) first produced protons when he bombarded hydrogen with alpha particles. Protons are emitted in artificial nuclear reactions and may likewise be produced when hydrogen ions are accelerated in particle accelerators.

The proton range is given by the relation

$$d_p = 1.77 \times 10^{-31} \times v_p^{3.9}, \tag{2.30}$$

where v_p is the velocity of protons [cm sec^{-1}].

Protons with an energy of 100 MeV have a range of a few meters in air at atmo spheric pressure. Protons with this energy occur also in cosmic radiation.

Proton energy is determined by means of a magnetic spectrograph, or on the basis of proton absorption by matter, or again, by measuring the length of the tracks they leave in a nuclear photographic emulsion.

2.4.1.6 Deuteron Radiation

One of the isotopes of hydrogen is deuterium (^2_1D) and its atomic nucleus is called deuteron. Deuteron radiation comes from artificial nuclear reactions, but may be produced also in deuterium-fed accelerators.

The range $d_{d(E)}$ [cm] of a deuteron radiation with an energy of E [MeV] is equal to twice that of proton $d_{p(E/2)}$ with an energy of $E/2$ [MeV]. Hence, the following relation is valid:

$$d_{d(E)} = 2d_{p(E/2)} \text{ or } d_{d(2E)} = 2d_{p(E)}. \tag{2.31}$$

2.4.2 Sources of Nuclear Radiation

Nuclear analytical chemistry makes use of the most diverse sources of nuclear radiation. They are utilized in activation analysis, in analytical methods based on the measurement of nuclear radiation absorption and backscatter in the preparation of radioactive indicators, etc. The sources of nuclear radiation made use of in nuclear analytical chemistry may be divided into three groups:

1. radionuclide sources,
2. accelerators,
3. nuclear reactors.

These three types of radiation sources will be briefly described in the present chapter. We shall concentrate above all on a description of the basic components that enter into the building and activity of these sources. More detailed data will be introduced later when individual analytical applications are treated.

2.4.2.1 Radionuclide Sources

Radionuclide sources of nuclear radiation comprise natural and artificial radioactive nuclides that emit alpha, beta and gamma radiations. Today about 50 natural radioactive nuclides are known and the number of those produced artificially runs into thousands. Hence, a very wide choice of radioactive elements is at our disposal, and for the solution of a certain analytical problem we may select a source which will prove most satisfactory from every aspect – not excluding the economic factor.

Analytical applications for alpha emitters include methods making use of alpha radiation absorption and backscatter, their powerful ionization ability, the use of activation analysis, etc. In the case of beta emitters practical applications utilize their transmission, absorption and backscatter: in addition, they are used as sources of X-rays. Gamma emitters are employed in analytical work involving absorption and backscatter of gamma radiation, further, for determining beryllium and deuterium by photoactivation.

Radioactive elements were the first sources of particles for bombarding atomic nuclei to produce nuclear reactions. The advantages of these sources lie in that they are relatively easily available, require little space and are simple to manipulate. But they also present a number of drawbacks, especially when employed for activation purposes (in activation analysis). They offer a very restricted range as regards the choice of projectiles (they are mostly useful as sources of alpha particles, gamma quanta), the energy of particles emitted by radioactive elements is only a few MeV and the radiation intensity is rather weak. For this reason they are used only as laboratory "radionuclide neutron sources."

Radionuclide neutron sources may be divided into two main groups according to the type of nuclear reaction in which neutrons are produced: (α, n) and (γ, n) neutrons source.

The condition for an (α, n) reaction is that the energy in the α particle be sufficiently great to overcome the potential nuclear barrier of the target material and that the reaction be exoergic. The most common sources of alpha particles are the natural alpha-active elements (Ra, Rn, Po and others), but artificial ones, such as Pu, Am are also employed. Target material consists of light elements, mainly Be (less frequently Li, B, because of their low cross-sections). The sources consist of a mixture of the two components. At the present time a considerable number of various neutron sources are produced, with a varying neutron flux. The cost of the sources is dependent primarly on the availability and the amount of the radionuclide used. The choice of the source is governed by the purpose and the conditions of the application.

Neutron sources (γ, n) are also called photoneutron sources. The energy of gamma radiation employed should exceed the bond strength of the neutron in the nucleus. In the case of stable nuclides the lowest neutron binding energy is in ^{9}Be (1.666 MeV), in D it is 2.226 MeV. The neutron source (γ, n) most commonly utilized consists of radionuclide ^{124}Sb and beryllium as target material. Further sources of gamma radiation currently employed are ^{24}Na, ^{56}Mn, ^{72}Gd, ^{88}Y, ^{140}La, MsTh and Ra. In addition to beryllium, D_2O also serves as target.

A complete neutron irradiation equipment from *The Reactor Experiments*, Inc. [4] is illustrated in Fig. 2.19.

Radionuclide neutron sources are utilized in activation analysis, in neutron absorption analysis, in analytical methods based on slowing down and scattering of neutron radiation (e.g., determination of hydrogen, humidity, etc.).

These and further sources will be described in more detail in Volume III of our monograph series, in the chapter dealing with activation analysis.

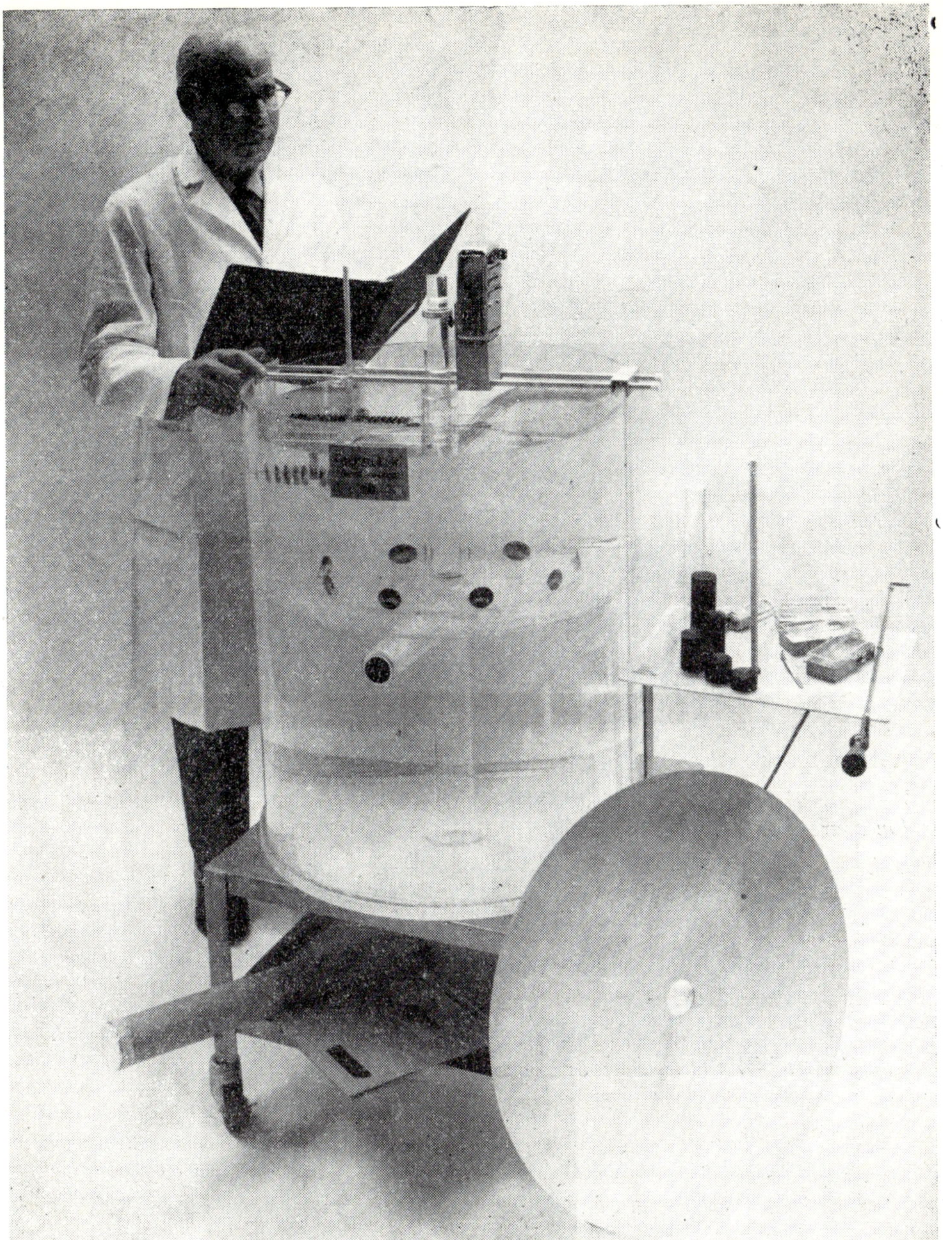

Fig. 2.19. Radionuclide neutron irradiation equipment, supplied by Reactor Experiments, Inc. (Courtesy of Reactor Experiments, Inc.)

2.4.2.2 Particle Accelerators

In order to reach the nucleus of an atom and achieve a nuclear reaction, a charged particle must approach sufficiently close to the nucleus to surmount the potential barrier of the nucleus. Hence, an important task of nuclear physics was to work out methods that would yield charged particles with energies of several million electronvolts. Today a large variety of devices are known by means of which electrons, protons, deuterons, or alpha particles of sufficiently high energy and corresponding intensity can be obtained. These devices are known as *particle accelerators*. Although they may be designed for special purposes they all work on the same principle. Electrically charged particles (ions) obtained in an *ion source*, are accelerated by means of an electric field; the movement of the particle may at the same time be affected by the action of the magnetic field. According to the shape of the pathway along which the particles are being accelerated, we may divide accelerators into two categories, viz.:

a) those in which particles are accelerated along a direct pathway (here belong acceleration tubes with voltage sources: multistage high-frequency accelerators and linear accelerators with wave conductors),

b) those in which particles are accelerated along a circular or a spiral path (cyclotron, synchrocyclotron, betatron, electron or proton synchrotron).

Accelerators consist of an electron or ion source, or a larger number of accelerating electrodes and a collector (target) on which the nuclear reactions are induced.

Ion sources used in accelerators are of various types. Among the most current belongs the arc maintained by the electron flux from an incandescent filament. In the high-frequency source, ions are produced by a discharge induced in a high-frequency magnetic field.

The early constructions of accelerators made use of a high-voltage induced into a so-called *accelerating tube*, to produce accelerated particles. The necessary voltage was supplied by a suitable high voltage generator (van de Graaff or cascade generator).

The *Cockroft-Walton generator*, used even now in various designs, and of significance also for analytical chemistry, was the first with which artificial nuclear transformation was achieved. This voltage accelerator gains through voltage doubling on the capacitor-rectifier system (Greinacher coupling). The resulting voltage at the ends of the series is the sum of these partial voltages on the capacitors. The electrical charge spent through its discharge in the acceleration tube is replaced by a new charge.

In the *van de Graaff generator* a continuous moving belt of a suitable insulator brings the charge from the high voltage source (*A*). The charge is removed through the terminal point and loads a hollow spherical electrode (*B*) (conductor), to the point of breakdown voltage of the surrounding gas. This breakdown voltage may be increased if the electrode is placed in the space with a gas of high electrical strength. A schema of the van de Graaff accelerator with the accelerating tube is shown in Fig. 2.20. Proton energies up to 5–7 MeV were obtained with the van de Graaff generator, which about represents the limit of energies attainable in one accelerating circuit.

Van de Graaff generators are widely used also in nuclear analytical chemistry, in activation analysis. Figure 2.21 shows a photo of a van de Graaff generator (Model AN 400 Positive Ion Accelerator) supplied by the High Voltage Engineering Corp., which has found application in activation analysis and offers wide prospects for various areas of nuclear analytical chemistry. Figure 2.22 shows this accelerator in an open state and with the regulating unit.

Today, the greatest significance is attached to accelerators in which the required energy is obtained gradually, that is by the repeated action of relatively small voltages.

The most common, and with respect to radiochemical purposes, the most significant ion accelerator is the *cyclotron.*

In this device a magnetic field is used to make particles move in a spiral of increasing radius, so that the length of the path automatically increases with the speed of the accelerated particles.

The cyclotron consists of two powerful magnets. Between them lies a hollow chamber in which are two semi-circular metal boxes, called dees. The electrodes are connected to the high-frequency voltage source and the particles are accelerated by the electrical field between the dees. The passage of the particles between the dees must be synchronized with the high-frequency accelerating field so that they pass in between at the very instant at which the field is able to accelerate them. Spatial stability is attained by means of magnetic focusing. The particle(ions) source is located in the center of the cyclotron and the beam of accelerated particles can be brought out of the dee chamber by means of a deflecting plate which is charged to a high negative potential. The attracting force acting on the positive ions draws them out of their spherical paths, so that they can be used to bombard any desired target. The principle of a cyclotron is represented in Fig. 2.23. Figure 2.24 shows a photograph of the cyclotron installed in the Institute for Nuclear Research ČSAV (Řež near Prague, ČSSR). Medium-sized cyclotrons are capable of accelerating protons, deuterons and helions up to energies of about 10 MeV.

In spite of the fact that large cyclotrons enable very high energies to be attained (the one installed at the University of California yields about 6500 MeV) their performance is essentially limited by the relativistic growth of the mass of revolving particles, which results in a prolonged transit through the dee; thus the particle would no longer be in phase with the oscillating potential. This problem has been solved by means of a frequency modulation of the accelerator, the so-called *synchrocyclotron.* Every ion impulse is driven by the cyclotron through a series of accelerating impulses

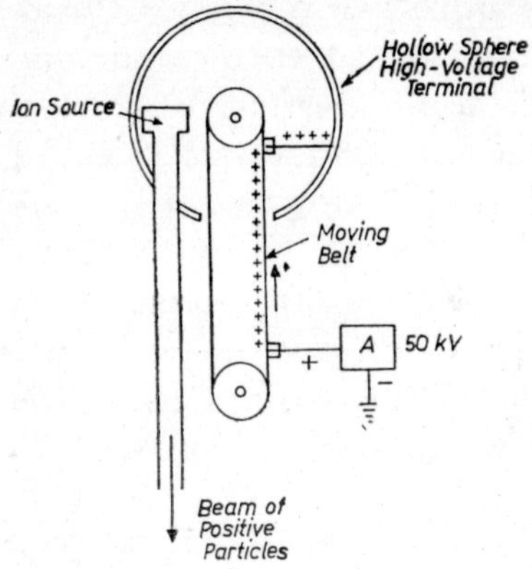

Fig. 2.20. Scheme of a van de Graaff generator with accelerating tube.

Fig. 2.21. 400 keV van de Graaff generator (Courtesy of High Voltage Engineering Corp.).

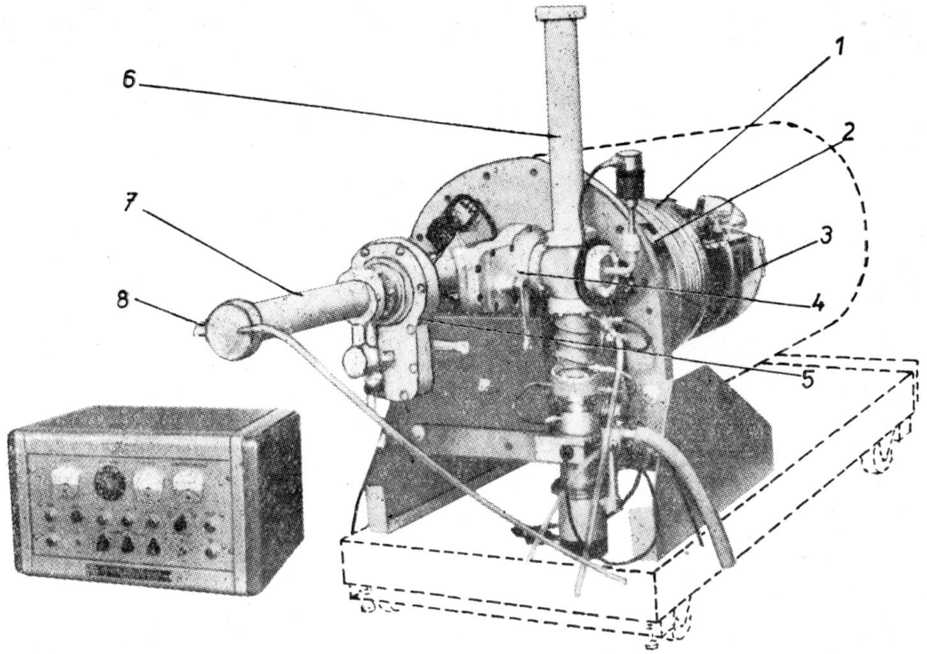

Fig. 2.22. Detailed view of a van de Graaff generator with control unit (1 — generator column, 2 — *r-f* ion source, 3 — high voltage terminal, 4, 5 — high-vacuum valves, 6 — extension system, 7 — accelerator tube, 8 — target).

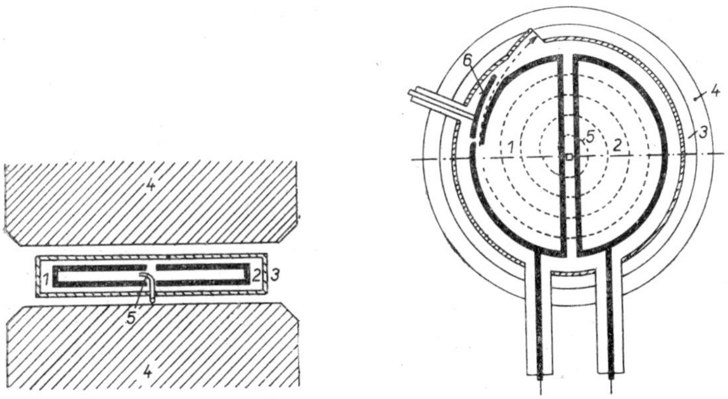

Fig. 2.23. Cyclotron (schematic) (1, 2 — pair of accelerating electrodes, 3 — vacuum chamber, 4 — electromagnet, 5 — ion source, 6 — deflection electrode).

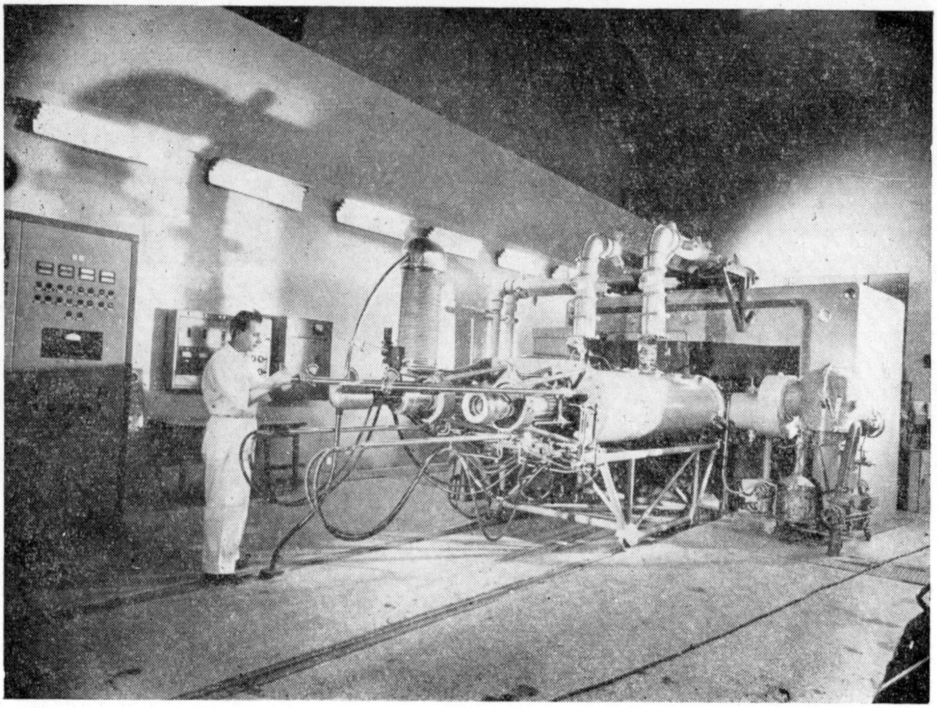

Fig. 2.24. Cyclotron installed at the Institute of Nuclear Research ČSAV (Řež near Prague, Czechoslovakia). Irradiation chamber, with electromagnet in the background. Energy of accelerated helium ions: 25 MeV.

with a moderately decreasing frequency so that the particles may be maintained longer in the field and may be imparted higher energies in a smaller installation.

Today, the so-called *betatron* is currently used for electron acceleration. Here, electrons revolve in the field of a strong electromagnet along a circular path and are inductively accelerated by oscillations in the magnetic flux.

Roughly speaking, a combination of the accelerating process of a betatron and a cyclotron gives rise to the so-called *synchrotron (synchrophasotron)*, which enables electrons as well as ions to be accelerated to extremely high energies. (The proton synchrotron CERN in Geneva generates an energy of 28,000 MeV.)

Accelerators will be treated in more detail when their applications are discussed.

Accelerators make it possible to produce nuclear missiles of all types, from those with the lowest energy to those whose energy equals many thousands of MeV. However, they cannot provide neutrons directly. These can be obtained in nuclear reactions induced when suitable targets are subjected to bombardment by accelerated ions or artificial gamma radiation. The targets used most frequently are ^{2}H (in the form of heavy ice, D_3PO_4, NaOD), ^{3}H, ^{7}Li, ^{9}Be, ^{12}C and ^{13}C. The most significant are the (*d*, *n*) reactions. Table 2.2 [5] lists the most important nuclear reactions for producing neutrons.

Table 2.2

Reaction	Q [MeV]
$^{2}H(d, n)^{3}He$	3.27
$^{3}H(d, n)^{4}He$	17.60
$^{9}Be(d, n)^{10}B$	3.79
$^{7}Li(d, n)^{8}Be$	15.00
$^{7}Li(p, n)^{7}Be$	—1.60
$^{2}H(\gamma, n)^{1}H$	—
$^{9}Be(\gamma, n)^{8}Be$	—

Neutron yields increase with the energy of the bombarding particles. The largest yield of neutrons can be obtained through the Be + *d* reaction, where yield is approximately 10^{10} n sec^{-1} for 1 μA of deuterons of an energy equivalent to 8 MeV. A given deuteron energy yields neutrons of maximum possible energy when the target is Li or ^{3}H (see the second column of Table 2.2).

2.4.2.3 Nuclear Reactors

Nuclear reactors are the most intense sources of slow and fast neutrons and of gamma radiation amounting to several MeV. They permit production of a large variety of nuclides to carry out neutron activation and absorption analyses, and they find a wide scope of application in other fields of nuclear analytical chemistry as well.

Nuclear reactors are devices in which the inner nuclear energy is released under controlled conditions during a chain fission (see p. 78) of the atomic nuclei of certain nuclides:

$$^{233}_{92}U,\ ^{235}_{92}U \text{ or } ^{239}_{94}Pu.$$

The design and construction of nuclear reactors differ according to the purpose of their application. Every nuclear reactor consists of two essential parts, shown in the schema of Fig. 2.25. The *active zone* (core of the reactor) is the inner part comprising the fission material, the so-called *nuclear fuel* (1), and the *moderator* of fast neutrons (2), most frequently consisting of graphite, heavy water, and ordinary water. The neutrons are slowed down in the moderator by means of elastic collisions. Those substances are selected as moderators that consist solely of atoms of low atomic number. The entire active zone is enclosed in a heavy metal *sheathing* (3).

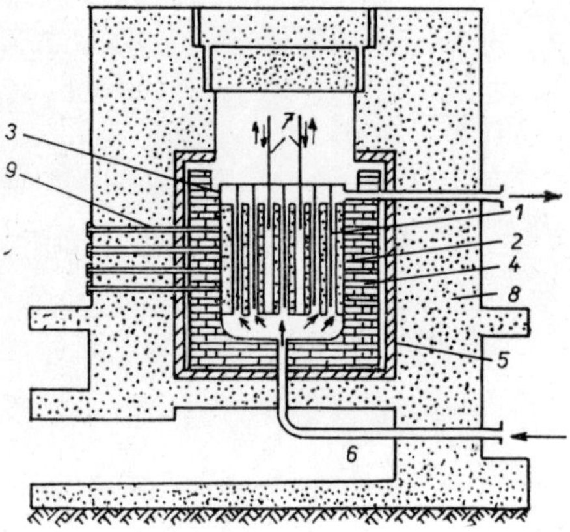

Fig. 2.25. Nuclear reactor (schematic).

To reduce neutron escape, the active zone is surrounded by a *reflector* (4) which deflects neutrons back into the active zone. Water, heavy water, beryllium, beryllium oxide or graphite may serve as reflector agents.

The thermal energy released during fission has to be removed by cooling the active zone. This *cooling system* consists of a series of tubes with an inlet and outlet tubing (6) through which the coolant circulates (gases, heavy and ordinary water, liquefied metals, etc.) The active zone, with the reflector and the cooling system, are located inside the *reactor container* (5). To bring the reactor up to its normal operating level and to maintain it at that level, a *regulatory system* consisting of control rods (7) of cadmium or of boron steel is used. These rods absorb all excess neutrons. The reactor is surrounded by a powerful *protective shielding layer* (8) through which holes or *irradiation channels* (9) permit access into the active zone. Samples to be activated (preparations of radionuclides, activation analyses) are introduced into these channels. The irradiation channels may be either horizontal or vertical.

In judging a reactor as a source of neutrons, the general law should be considered that the neutron flux is proportional to its energy level, the proportionality constant depending on the type and design of the reactor, and that it diminishes from the center outwards.

The neutron energy distribution likewise depends upon the type of reactor, for instance, thermal neutrons predominate in the uranium-graphite reactors, though fast neutrons, with an energy up to several MeV, cannot be neglected.

Nuclear reactors may be classified according to several aspects.

In *considering the neutron energy* used for fission, we recognize:

Thermal reactors (fission is induced by thermal neutrons), *fast reactors* (without a moderator; fissions are caused mainly by neutrons of energies above 0.1 MeV) and *intermediate reactors* (fissions are caused by medium energy neutrons).

According to the type of the active zone, reactors may be *homogeneous* (nuclear fuel and moderator form a homogeneous mixture), or *heterogeneous* (may involve solid fuel and solid moderator, or solid fuel and liquid moderator: the combination of liquid fuel with a solid moderator is also possible). Figure 2.25 illustrates the principle on which a heterogeneous reactor is built.

According to their *principal purpose* we classify reactors as *research* and *power reactors*, and further, as *reactors producing plutonium, radionuclides*, etc.

Reactors can be classified also according to the *type of fuel*, for instance, as *reactors with natural* or *enriched uranium* (with ^{235}U isotopes), according to the *type of moderator* as *graphite reactors, reactors with ordinary* or *with heavy water*, etc. Further, we recognize generating or *breeder reactors* in which reactions of neutrons produce secondary fuel (^{239}Pu).

An example of a research water-cooled reactor with enriched uranium is that installed in the Institute for Nuclear Research ČSAV (Řež near Prague, ČSSR), shown in Fig. 2.26. Because of their relatively low cost, TRIGA reactors [6–8] (supplied by the General Atomic Division, General Dynamics Corp.) are very suitable for purposes of activation analysis and other nuclear analytical applications. Reactor TRIGA Mark 1 (Fig. 2.27) performs at a stable rate of 250 kW and yields a neutron flux of 1.8×10^{12} n cm^{-2} sec^{-1}. More detailed features of this reactor will be given in the appropriate section dealing with activation analysis.

2.4.2.4 Recovery of Neutrons of Constant Velocity

Because of the high selectivity in neutron absorption, i.e., on account of the existence of resonance phenomena (see p. 74), in certain analytical applications (e.g., in neutron absorption analysis) neutrons of a constant velocity (monoenergic neutron flux) are required. Several methods for neutron monochromatization are known. We shall mention only the more important among them.

2.4.2.4.1 Mechanical Neutron Selector

In the first neutron selector, set up as early as 1935 [9], the neutrons were selected mechanically. The principle of this device is the following (see Fig. 2.28). Four duralumin discs are placed in the path of a narrow beam of thermal neutrons;

Fig. 2.26. Research reactor of the Institute of Nuclear Research ČSAV (Řež near Prague, Czechoslovakia) with a thermal output of 2 MW.

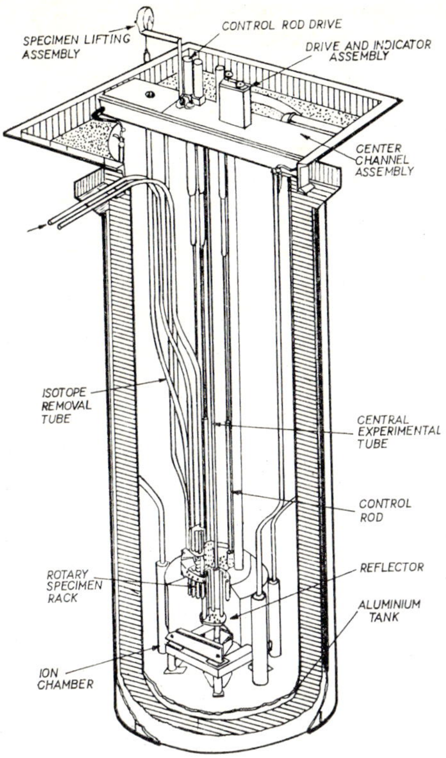

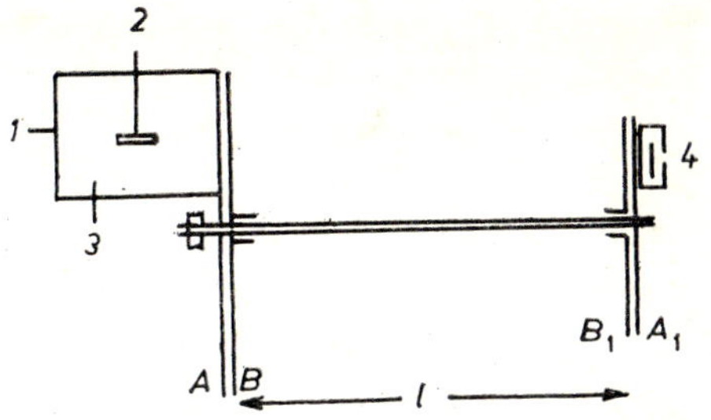

Fig. 2.28. Mechanical neutron selector (A and A_1 — fixed duralumin discs, B and B_1 — revolving duralumin discs, 1 — protective cadmium can, 2 — neutron source, 3 — paraffin, 4 — ionization chamber).

Fig. 2.27. Reactor TRIGA Mark I (Courtesy of General Atomic Division, General Dynamic Corp.).

two of the discs (A and A_1) are fixed; the other two (B and B_1) revolve rapidly. Each carries 50 cadmium sectors with duralumin gaps in between. As duralumin is practically transparent to neutrons, while cadmium absorbs them readily, neutrons will be captured whenever the position of the revolving cadmium sectors agrees with that of the duralumin gaps; when the cadmium sectors are mutually opposite, neutrons pass readily through the duralumin gaps. Such a system of synchronously revolving discs is represented by the monochromator, which allows neutrons to pass through at a speed of $v = l/t$, where l is the distance between discs B and B_1 and t is the time interval needed for a neutron to traverse this distance. A change in the number of revolutions, in the width of the cadmium sector and in the distance between the revolving discs can separate only neutrons of a certain speed.

2.4.2.4.2 Selectors with a Cyclotron-Modulated Beam

If the neutron source has a constant radiation intensity, we shall use the method of mechanical velocity selector. In the mechanical selector (Fig. 2.28) the left pair of discs modulates the intensity of the neutron flux. This modulation can be achieved by means of a method (its principle will be described later) which is at the basis of the construction of modern selectors with a cyclotron-modulated beam.

The neutron source is supplied by a beryllium target which is bombarded by protons or deuterons from the cyclotron. By modulating the current or the voltage of the

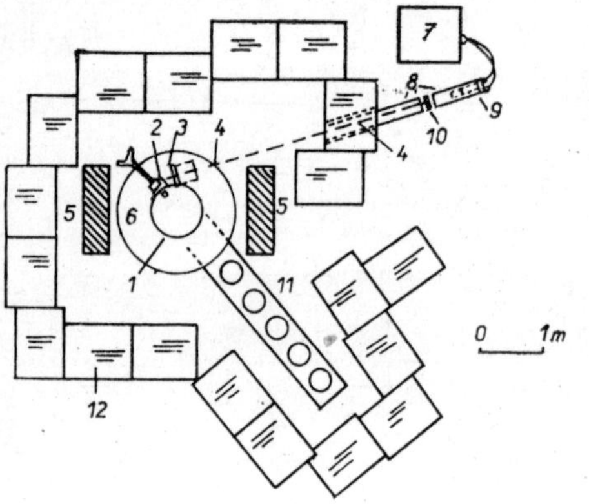

Fig. 2.29. General scheme of the selector with cyclotron-modulated beam (1 — cyclotron chamber, 2 — beryllium target, 3 — paraffin source of slow neutrons, 4 — collimator, 5 — electromagnet core of cyclotron, 6 — exciting coil, 7 — amplifier, 8 — collimator, 9 — counter with BF_3, 10 — sample, 11 — resonant line with dee lines, 12 — vats with tub).

electric arc that produces the ions, later accelerated in the cyclotron, we can effect the required periodic changes of the neutron flux intensity. If the entry of ions into the accelerating system is stopped, then the impact of the accelerated ions on the target will also be interrupted and the neutron production in the target can be modulated. A cyclotron is then a periodic source of neutrons and a neutron detector at a distance l from the target, capable of working for a short period of time (only after the lapse of an interval t following the bombardment charge) and of recording neutrons with $v = l/t$. As the cyclotron target emits only fast neutrons and the selector works with slow ones, the former have to be slowed down with a paraffin layer which is then a direct source of neutrons for the selector. The recording system consists of an ionization chamber filled with ^{9}B enriched boron fluoride or chloride, or a proportional counter containing the same substance and a multi-channel analyzer. The channels in the analyzer record neutrons with the corresponding velocity.

Figure 2.29 represents a schematized view of a selector [10]. The neutron flux is collimated with paraffin blocks, boron or cadmium. The sample to be analyzed is inserted between the neutron source and the detector.

2.4.2.4.3 Crystal Monochromator

Another possible way to obtain a monochromatic neutron beam is to decompose the neutron flux on a crystal into an energy spectrum, in a process similar to that taking place in X-ray diffraction.

If a narrow beam of neutrons is directed to a monocrystal, neutrons will be deflected at an angle α, which satisfies the condition

$$2d \sin \alpha = n\lambda, \tag{2.32}$$

where d is the spacing of the reflecting planes in the crystal, λ de Broglie's

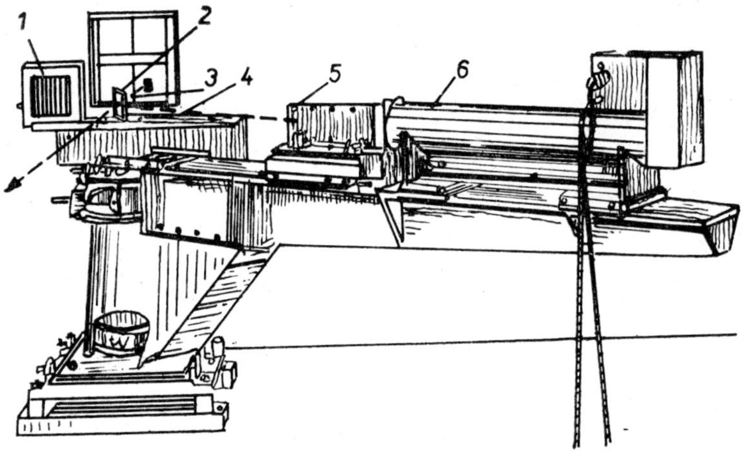

Fig. 2.30. Crystal monochromator of neutrons (1 — screening of neutron beam, 2 — crystal, 3 — incident beam, 4 — scattered neutron beam, 5 — parallel slits, 6 — BF_3-filled proportional counter).

wavelength, and n a positive integer. With crystal LiF, often used for neutron diffraction, $d = 2.32$ Å. Thus, with an angle $\alpha = 3.5°$, neutrons are emitted with a kinetic energy of 1 eV. (De Broglie's wavelength of these neutrons is 0.287 Å.) Neutrons of an energy equal to 100 eV are eliminated at an angle $\alpha = 0.35°$. An energy of 100 eV is practically the highest energy which can be eliminated by this method, for the angle divergence of the incident beam is of the same order.

The neutron source for a crystal monochromator is generally a nuclear reactor. A narrow beam of neutrons escaping through a slit in the reactor wall strikes the monochromator crystal and the reflected monochromatic neutron beam in its turn strikes the sample.

Figure 2.30 illustrates a typical crystal monochromator.

2.5 Interaction of Nuclear Radiation with Matter

When particles (α, β^+, β^-, n, etc.) or gamma photons pass through a substance, interactions occur with the matter. By the interactions the energy of the radiations might be reduced and the direction of the particle beams might be altered.

We shall now describe the laws according to which the interaction between various types of nuclear radiation and the substance takes place.

2.5.1 Alpha Radiation

Alpha particles, in their interaction with matter, lose energy. When passing through matter, their pathway is generally a long straight line. They lose their energy gradually until the instant they are suddenly stopped and ejected from their path. The interaction of alpha radiation with absorbing medium takes place especially in the electron shell of the atoms. This interaction may lead to *ionization* or *excitation* of the atoms and molecules. The alpha particles take up one or two electrons from the medium and are converted to unipositive helium ions (He^+), or to neutral helium atoms. However, because of the high velocity of the moving particles, the captured electrons split off rapidly and the process is repeated several times. At the same time ions with an opposite charge are produced from the molecules or atoms of the absorbing medium. Some of the electrons ejected by alpha particles acquire such a high kinetic energy that they themselves can produce ions. Alpha particles moving in a gas medium produce a large number of ions in their pathway. For example, in air at 15 °C and at a pressure of 750 mm Hg, one alpha particle creates from 150 to 200 thousand ion pairs. The number of ion pairs produced per unit length of the trajectory of the alpha particle *(specific ionization)* gradually increases and attains a maximum at the end of its path (Fig. 2.31). For each ion pair formed in air an average of 32.5 eV are dissipated. If the energy transferred from the alpha particle to the atom is less than 32.5 eV, only excitation of the atom or molecule may occur. When the atom or molecule loses its excess energy, light quanta are emitted. The atom which emitted the alpha particle is reflected and acquires kinetic energy of the order of 0.1 MeV. This energy is sufficient for the particle produced to manifest a certain ionization capacity.

The range of alpha particles is indicated, in particular, by the rate of energy loss due to ionization caused by the alpha particle in the medium through which it passed. The range increases with increasing alpha energy. The relationship between the range and the initial velocity of alpha particles can be expressed by *Geiger's law*

$$R = k \cdot v_0^3, \tag{2.33}$$

where R is the range of alpha particles [cm], v_0 the initial velocity [cm sec^{-1}] and k a constant (equal to 9.25×10^{-28}) [sec^3 cm^{-2}].

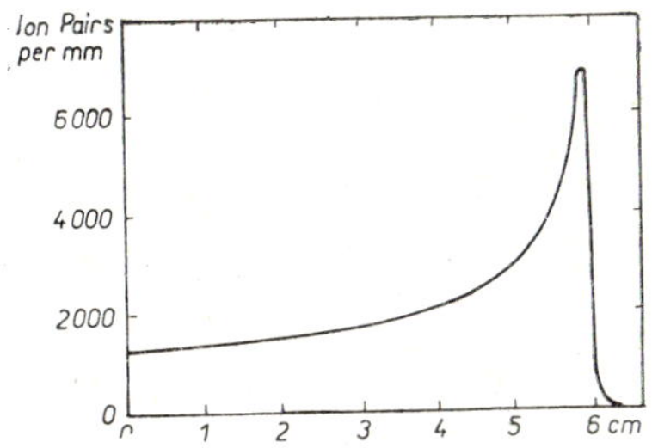

Fig. 2.31. Correlation between specific ionization and distance from source.

All alpha particles emitted by an atom possess approximately the same energy, i.e., the ranges of alpha particles emitted by a given radionuclide differ very little and fluctuate, according to a statistical curve, about the mean value.

The connection between the range of alpha particles and the decay constant is expressed by the *Geiger-Nutall relation*

$$\log \lambda = A \log R + B, \tag{2.34}$$

where A is a constant, and has the same value for all natural alpha-emitting nuclides, and B a constant for alpha-emitters within a given decay series. This relation is only approximate and a major disagreement is found mainly in UI, AcX and short-term C′, products in all three series.

The range of alpha particles decreases with increasing atomic weight and atomic number of the absorbing element. In substances other than air, the range may be calculated with a precision of $\pm 15\,\%$ by the *Bragg-Kleeman rule*

$$R = 0.0003 \frac{R_0}{\varrho} A^{\frac{1}{2}}, \tag{2.35}$$

where R_0 is the range of alpha particles in air, and R the range of the given substance of density ϱ.

Alpha particles are, rarely, *scattered* by the electrostatic field of the nucleus, or reflected (one in 10^4 alpha particles), or a *nuclear reaction* takes place (one in 10^6 alpha particles).

The interaction processes of the other types of ionizing radiations (protons, deuterons, light ions, fission products) are qualitatively similar to that described for alpha radiation, though the mechanism of interaction may differ.

2.5.2 Beta Radiation

The processes which are responsible for the energy loss in beta radiation are in many ways fundamentally similar to those described for alpha particles and other ions. The basic forms of energy transfer in beta radiation are *ionization* and *excitation.* Interactions of beta radiation with atomic nuclei leading to nuclear reactions have so far been observed only in the case of high energy radiation.

The gradual slowing-down of beta particles in the absorbing medium is due mainly to its interaction with the electrons of the atom. At about 1 MeV of energy this interaction leads to ionization. The beta particles are much lighter than alpha particles; for a given energy their velocity is larger than that of alpha particles and therefore the specific ionization is less for electrons. The range of beta particles is greater than that of alpha particles. In air a 1 MeV beta particle can produce approximately 30,000 ion pairs. The specific ionization of beta radiation in air at 15 °C and 760 mm Hg is given by the relation

$$\text{Specif. ionization} = \frac{45}{\beta^2}, \tag{2.36}$$

where $\beta = v/c$ represents the relative velocity of the beta particles in relation to the speed of light. The absorption of beta particles is substantially more complex than that of alpha radiation: their spectrum is continuous and they may lose a large part of their energy in one collision. Hence, beta particles are considerably scattered in their path.

Because of the continuous nature of beta radiation spectrum and the scattering of beta particles, beta radiation cannot be characterized by a constant range for all the particles. They are far more penetrating than alpha particles. The beta range in air may vary from fractions of a millimeter up to several meters. Penetration by beta particles in a substance depends above all on their initial energy. A constant range is observed only for monochromatic radiation. In a given element this may apply solely to beta particles with the highest energy. The thickness of a medium absorbing beta particles with maximum energy is termed *total absorber thickness.*

Energy decrease in beta particles through ionization is practically proportional to the product $N \cdot Z$ (where N is the number of atoms in 1 cm^3 and Z the atomic number of the absorber element). The specific ionization of low energy beta particles is relatively large while their range is short. With increasing beta particle energies, however, this specific ionization rapidly decreases to a minimum value, corresponding approximately to a 1 MeV particle energy, and thereafter it remains practically constant. If the energy of beta particles is insufficient to produce ionization, the atoms or molecules are excited.

When beta particles of higher energies pass through an absorber, an electromagnetic radiation, known as "*Bremsstrahlung*," takes place. The intensity of this radiation depends on the atomic number of the absorber. The origin of "Bremsstrahlung" is similar to that of an X-ray when high energy electrons hit a heavy metal target. Maximum energy of "Bremsstrahlung" equals that of the beta particles: here, however, the mean energy is much lower than the maximum. The origin of the so-called *annihilation radiation* and the *characteristic X-rays* during interaction with beta radiation have been described above.

We have mentioned that beta radiation may be absorbed in various ways during the interaction processes. The absorption of beta particles is represented by an exponential law:

$$I = I_0 \cdot e^{-\mu' d}, \tag{2.37}$$

where I_0 is the measured intensity without absorber, I the intensity observed through absorber of thickness d, and μ' the linear absorption coefficient [cm^{-1}].

In dividing the linear absorption coefficient (μ') by the density of the absorber (ϱ), we obtain the mass absorption coefficient (μ) [$cm^2\ g^{-1}$], and then equation (2.37) may be written as follows:

$$I = I_0 \cdot e^{-(\mu'/\varrho) R} = I_0 \cdot e^{-\mu R}, \tag{2.38}$$

where R is the square weight of the absorber [$g\ cm^{-2}$].

In characterizing the absorption of beta radiation we often give that thickness of the absorber medium which reduces the intensity of the incident radiation to half its value, termed also *half-thickness* ($d_{\frac{1}{2}}$). Hence, for $d = d_{\frac{1}{2}}$, $I = I_0/2$, and if these values are substituted in equation (2.37), then

$$d_{\frac{1}{2}} = \frac{0.693}{\mu'} \quad [cm], \tag{2.39}$$

or from equation (2.38) we obtain

$$R_{\frac{1}{2}} = \frac{0.693\varrho}{\mu} \quad [g\ cm^{-2}]. \tag{2.40}$$

2.5.3 Gamma Radiation

Gamma radiation is the most penetrating type of radioactive radiation. Its range in air amounts to several hundred meters, in solids from a few centimeters to several decimeters, depending on the density of the absorber and the energy of the gamma quanta. The processes causing energy losses in gamma radiation differ essentially from those involved in corpuscular radiation. Gamma radiations produce ionization entirely by transferring their energies (totally or partially) to electrons;

thus ionization is secondary in nature. Three processes are important by which gamma radiations lose their energy: the photoelectric effect, Compton scattering and pair production.

The photoelectric effect is the most important phenomenon when low energy gamma quanta and high atomic number of the absorber are involved. In this process the whole gamma quantum energy ($h\nu$) is transferred to a bound electron

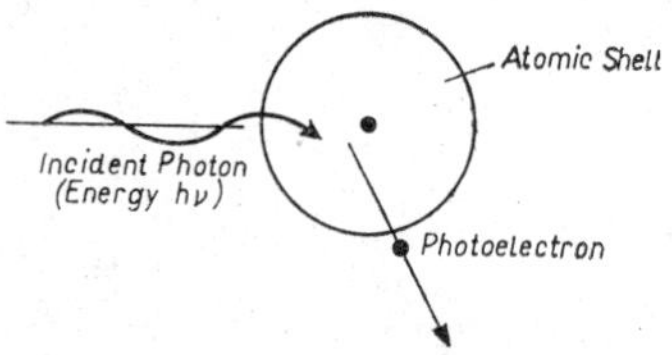

Fig. 2.32. Photoelectric effect.

and this is ejected from an atom or molecule (Fig. 2.32). Thus the electron acquires a kinetic energy of E

$$E = h\nu - E_v, \tag{2.41}$$

where E_v is the binding energy of the electron, while the gamma quantum completely disappears.

Following the ejection of an electron, the electronic level is immediately restored and a characteristic X-radiation can be observed. The energy of this X-radiation is often immediately transferred to a more loosely bound outer electron which then quits the atom. This is known as Auger electron. Photoelectrons lose their energy through ionization of the "Bremsstrahlung" processes in the form of beta radiation.

Compton scattering is observable at medium gamma energies (up to 2 or 3 MeV) and in elements with a low atomic number (up to $Z = 13$). During the course of this phenomenon an elastic interaction takes place between the gamma quanta and some outer electron of the atom in the absorber medium, or with a free electron. Valence electrons are loosely bound and therefore in collision they behave practically as free electrons. The incident gamma quantum transfers a part ($h\nu'$) of its energy ($h\nu$) to an electron which is ejected with an angle of φ in relation to the original path of the photon. The gamma quantum is degraded in energy $h\nu'' = h\nu - h\nu'$ and is deflected with an angle of α from its original path (Fig. 2.33). The scattered gamma quanta lose their energy in their interaction with electrons of the atom and their absorptions result in photoelectric effects. The change in wavelength $\Delta\lambda$ may be calculated according to the following equation

$$\lambda' - \lambda = \frac{h}{m_0 \cdot c}(1 - \cos \alpha) = 0.242(1 - \cos \alpha), \tag{2.42}$$

where m_0 is the rest mass of the electron.

The maximum energy of Compton's electron may be calculated by the formula

$$E_{\text{Compt}} = \frac{E_\gamma}{1 + \frac{0.511}{2E_\gamma}}, \tag{2.43}$$

where E_γ and E_{Compt} are given in MeV.

Pair production is a process which involves the creation of a positron–electron pair by a gamma quantum of at least 1.02 MeV. During this process in the field

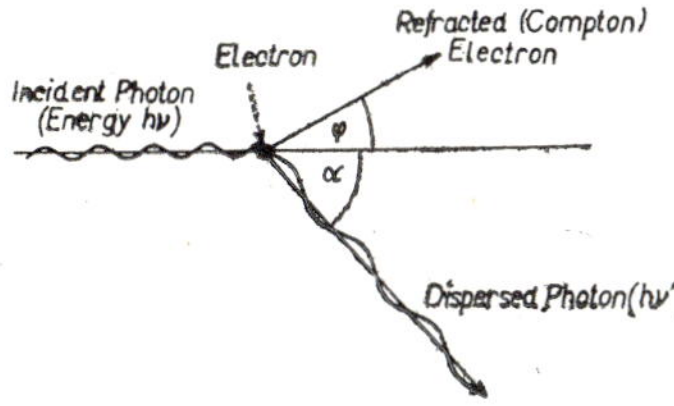

Fig. 2.33. Compton scattering.

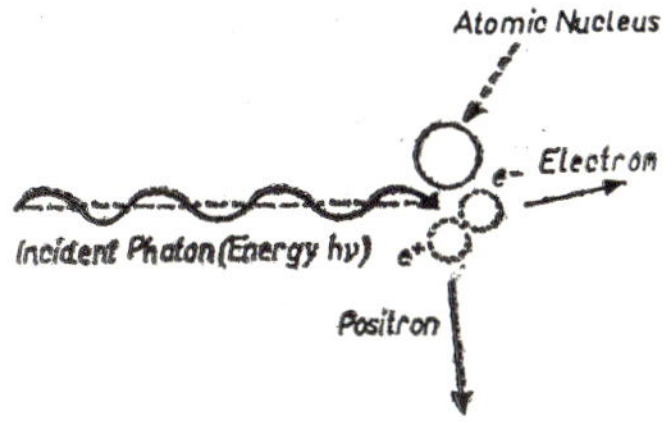

Fig. 2.34. Electron pair production.

of a nucleus, gamma quantum is converted into a electron–positron pair (Fig. 2.34). The entire energy of the gamma quantum is used up in this process, partly in pair formation and partly in imparting kinetic energy. As the rest mass of the electron and positron corresponds to 0.51 MeV, the electron–positron pair is formed at a certain threshold energy which is the sum of the rest mass of both the particles (1.02 MeV), and the excess energy of the gamma quantum is converted into their kinetic energy. Pair production is always followed by the annihilation of the positron, with the simultaneous emission of two 0.51 MeV gamma quanta.

In addition to these processes, the so-called *classical deflection* may occur (though with very little probability) with low energy gamma quanta (< 0.02 MeV), when only the course of the photons becomes altered. With high energy gamma quanta (> 8 MeV), the so-called *nuclear photo-effect* takes place. Gamma quanta enter directly into reaction with atomic nuclei, thus giving rise to nuclear reaction.

During the passage of gamma quanta through a substance, their intensity gradually decreases because of the absorption processes and approaches asymptotically to zero. In contrast to beta radiation, there is no maximum range value here. Gamma absorption, like that of beta radiation, can be described by an exponential law (2.37). The absorption coefficient μ' in this equation (2.37) may be considered as a total absorption coefficient, which is the sum of three separate coefficients characteristic of the photoelectric effect τ, Compton scattering σ and pair production $\varkappa$

$$\mu' = \tau + \sigma + \varkappa. \tag{2.44}$$

The correlation between these partial absorption coefficients and the gamma radiation energy is given in Fig. 2.35.

If the total mass absorption coefficient is applied, equation (2.38) like the one in the case of beta radiation. However, the total mass absorption coefficient μ consists of three partial mass absorption coefficients, viz., that for the photoelectric effect (τ/ϱ), Compton scattering (σ/ϱ) and pair production $(\varkappa/\varrho)$.

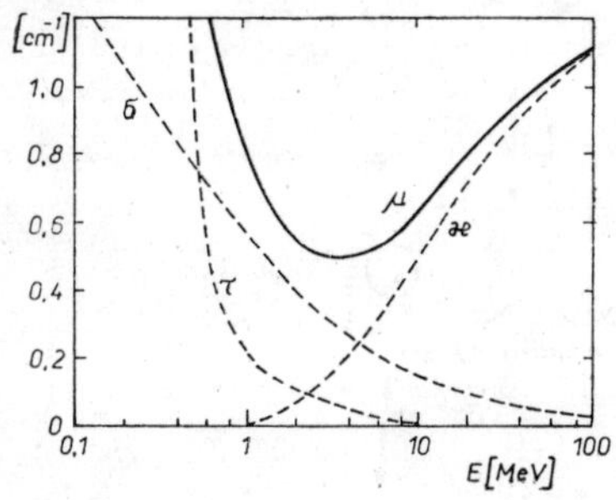

Fig. 2.35. Total and partial absorption coefficients for gamma radiation in lead.

The value for the half-thickness may be calculated from the relationships valid for beta radiation, i.e., according to relations (2.39) and (2.40).

Ionizing collisions of gamma radiation are caused by secondary electrons, which in their passage through the substance strike gamma quanta out of the atoms. A gamma quantum produces the same number of ions as an alpha or a beta particle, of the same energy: however, as gamma radiation is more penetrating, the ions produced will be scattered over greater distances. The number of ions grows rapidly with increasing absorption of gamma quanta.

2.5.4 Neutron Radiation

The character of the interaction between neutrons and a substance is determined by forces acting between the neutrons and the particles of the substance – electrons and atomic nuclei.

These forces are of very low energies. Neutrons have no electrical charge; hence, the Coulomb forces are not involved: since, however, both particles possess a magnetic momentum, an electromagnetic interaction between them cannot be ruled out. A magnetic interaction between these two particles becomes manifest only in special experiments with scattering of neutrons in ferromagnetic and paramagnetic substances. In these cases, in consequence of a weak but synergetic action of many electrons of the substance on the neutron, a certain coherent scattering occurs. A magnetic interaction exists also between neutrons and atomic nuclei. Nevertheless, this interaction is substantially lower than in the case of electrons, and in comparison with nuclear interaction, it is quite negligible. So far, these interactions have found no analytical application at all.

The behavior of neutrons in their passage through a substance is then determined nearly exclusively by their interaction with atomic nuclei. This interaction results from nuclear forces. The probability of an interaction's taking place between a nucleus and a neutron is substantially higher with thermal than with fast neutrons. The explanation advanced for this is that the lower the energy and, consequently, the velocity of the neutrons, the longer will they "stay" in the vicinity of the nucleus and hence, the greater probability of an interaction. Therefore, absorption analyses are carried out mostly with slow neutrons.

A collision between a neutron and a nucleus may result from a simple deflection of the original path of the neutron through the action of nuclear forces, i.e., by scattering or even capture of the neutron by the nucleus. An interaction between a neutron and a nucleus may take place as:

1. an elastic scattering,
   a) potential,
   b) resonance;
2. an inelastic scattering;
3. a nuclear reaction,
   a) ordinary nuclear reaction,
   b) fission,
   c) radiation capture (n, γ),
   d) spallation.

All the processes except elastic potential scattering are the results of neutron capture by the nucleus.

In an *elastic scattering*, a neutron transmits part or all of its kinetic energy to the nucleus of the atom in the absorber, but the form of the energy is not altered in the process. The sum of the kinetic energies before and after collision (of neutron with nucleus) remains the same. If a *central elastic collision* is assumed, then the neutron energy following collision (E'_n) will be

$$E'_n = E_n \left(\frac{A - 1}{A + 1} \right)^2, \tag{2.45}$$

where E_n is the neutron energy before collision, and A the mass number of the nuclide on which scattering took place.

In a neutron collision with heavy atomic nuclei the energy that is transferred is negligible ($A \gg 1$, hence $E'_n \doteq E_n$); that is, neutrons are scattered without their energy being appreciably reduced. If, on the contrary, we assume that $A = 1, E'_n = 0$, then the neutron loses almost all its kinetic energy. This condition is practically fulfilled when neutrons scatter in a medium containing hydrogen, for a neutron and a proton have nearly the same mass. In its central collision with deuterium a neutron transfers about 89 %, with carbon about 28 %, and with lead about 2 % of its kinetic energy. If a *non-central elastic collision* is involved, the neutron energy loss is smaller.

Elastic *potential scattering* takes place when the neutron energy is higher or lower than that of the resonance level. In a potential scattering the incident neutron does not really penetrate the target nucleus and therefore no compound nucleus is formed (see p. 70). A potential scattering is the simplest form of interaction of a neutron with a nucleus. This implies a deflection of the neutron from its original path under the action of nuclear forces.

Resonance scattering occurs when the neutron energy is close to or corresponds to the energy of one of the quantum states of a compound nucleus. With the exception of the region close to the resonance area, resonance scattering is always much smaller than the potential one. The elastic resonance scattering does not differ dynamically from the potential scattering. However, the probability of these two phenomena may vary and may be variously dependent on the kinetic energy of neutrons.

A characteristic feature of *inelastic scattering* is that part of the kinetic energy taken up by the nucleus from the neutron is transferred to the state of compound nucleus and only the remainder represents an increment of kinetic energy to the nucleus. The latter then gets rid of the excess energy by emitting gamma quanta. As a rule, the incident and the scattered electrons in an inelastic scattering are not identical.

The principal form by which neutron energy is transferred to an absorber of light atoms is the elastic collision. In absorber consisting of heavy atoms the decisive form of energy loss by neutrons is through inelastic scattering. Nuclear reactions with neutrons will be treated in more detail in Chapt. 2.6.

All the interaction processes of neutrons with atomic nuclei of the absorber lead to an attenuation of the neutron flux. This may be expressed by the following exponential relation

$$\Phi = \Phi_0 \cdot e^{-\sigma d N}, \tag{2.46}$$

where Φ_0 represents the flux of the incident and Φ that of the residual neutrons [n cm^{-2} sec^{-1}] after their passage through an absorber of thickness d, N is the number of atoms per cubic cm and σ is the total cross-section.

From among the absorption processes *radiation capture* is the most important for analytical applications, i.e., the capture of neutrons followed by gamma radiation. Less common reactions in which absorption of slow neutrons is involved are (n, p), (n, α), and nuclear fission. Resonance scattering, in which a neutron is first captured by a nucleus to form a compound nucleus and then is emitted, may also be considered as a nuclear reaction (n, n'). Resonance scattering of slow neutrons is a relatively rare phenomenon and its significance for analytical purposes has so far been negligible. No compound nuclei are formed in a potential scattering but the neutron is only deflected from its original path under the action of nuclear forces. The differences in the cross-sections for scattering of individual elements are very small. A reduction in energy brought about by the scattering effects may be evaluated only in the case of hydrogen. The greater the energy loss in various collisions, the smaller the nucleus involved in the neutron collision. Hence, chemical analysis may also utilize changes in neutron energy that occur in potential scattering. This principle can be applied in practice in the case of hydrogen.

2.6 Nuclear Reactions

So far we have concentrated mainly on nuclear transformations that take place spontaneously—radioactive conversions. However, the number of reactions that can be induced artificially in nuclear reactions is considerably higher.

The basic requirement for a nuclear reaction to occur is collision between the target nucleus (X) and the bombarding particle (a). The resulting atomic nucleus (Y) may be either stable or radioactive. If radioactive nuclei are formed, we may speak of *activation.*

In a nuclear reaction a particle or gamma quantum (b) is simultaneously emitted. A simple type of nuclear reaction may be generally expressed by the schema

$$\mathrm{X} + a \rightarrow \mathrm{Y} + b \tag{2.47}$$

or in an abbreviated form as

$$\mathrm{X}(a, b)\,\mathrm{Y} \tag{2.48}$$

or else reaction (a, b) on nucleus X.

Like every physical or chemical process, a nuclear reaction is connected with energy change.

If energy is to be supplied to particle a and nucleus X to produce system Y + b, the reaction is termed *endoergic*. If, on the contrary, energy is released in the process, the reaction is called *exoergic*.

If rest masses are indicated by the symbol M with the corresponding index (in energy units) the reaction under (2.48) may be rewritten as

$$M_{\mathrm{X}} + M_a = M_{\mathrm{Y}} + M_b + Q, \tag{2.49}$$

where Q indicates the energy of the reaction.

If this relation yields a negative value for Q, energy must be supplied in order that the system Y + b might be produced (endoergic reaction): in the opposite case, an exoergic reaction is involved.

2.6.1 Compound Nucleus

Nucleons are mutually firmly bound in the nucleus. A measure of this bond is the mean binding energy per nucleon (in the case of nuclei with a medium or higher atomic number this is approximately 8 MeV).

If the particle incident upon the nucleus has a kinetic energy lower than the mean binding energy per nucleon, the interaction of the incident particle may be considered as that between it and the whole nucleus. Under such conditions the model of the compound nucleus is employed as suggested, independently of one another, by Bohr [11], and Breit and Wigner [12]. According to this theory it is assumed that nuclear reactions take place in two separate processes:

a) The incident particle is attached to the nucleus and forms a *compound nucleus.* Schematically: Target nucleus + incidental particle (e.g., neutron) → compound nucleus. The particle swallowed by the nucleus supplies the atomic nucleus with excitation energy which is equal to the sum of the kinetic energy of the particle and its binding energy in the compound nucleus.

b) The compound nucleus is decomposed and emits a particle or a photon. Schematically: Compound nucleus → residual nucleus + emitted particle. This process is independent of the first, but depends solely on the properties of the compound nucleus. In the second stage of the nuclear reaction the compound nucleus loses excess energy. This process may run via inelastic scattering, exchange nuclear reaction, capture reaction, fission reaction, or spallation reaction.

Let us consider the case of interaction between a neutron and the target nucleus.

When a neutron approaches the nucleus to a distance at which the nuclear forces become operable, it enters into an interaction with the nearest nuclear particles and transmits to them a substantial part of its energy. The energy of the incident neutron is shared among the nucleons before any particle can be emitted. This energy-sharing is governed by statistical laws. The newly formed nucleus is then in an excited state and its excitation energy is divided among a large number of particles. At the same time, none of the particles possesses an energy sufficient for it to leave the nucleus. A particle may be emitted from an excited nucleus only under the effect of a fluctuation in the distribution of the excess energy transmitted to the nucleus by the neutron incident on the nucleons. If through such fluctuations adequate energy is accumulated on one particle for its escape, it will leave the compound nucleus. This may give rise to three enventualities:

a) the escaping particle will be identical with the incident one (neutron),

b) the escaping particle will be different from it (e.g., proton),

c) the neutron may be captured by the nucleus and the latter will emit the excess energy in the form of a photon.

In the case of light nuclei ($A < 10$) the compound nucleus theory is valid up to an energy of 20 MeV; for nuclei of a medium mass, up to 50 MeV; and for heavy nuclei, up to 80 MeV.

The stability of the excited nucleus may be expressed by its *lifetime* τ. This is the time during which the nucleus persists in a certain excited state before decaying. To every quantum state of a nucleus there corresponds a certain *level width* Γ (an energy level) which may be taken as the measure of the uncertainty of energy value corresponding to the given state. The width of the level is proportional to the probability of decay during a unit time which may take place under a certain energy state of the compound nucleus. The magnitude

$$\Gamma = \frac{h}{2\pi\tau} \tag{2.50}$$

has been selected as the measure of the level width; this magnitude has an energy dimension and is usually expressed in electron-volts (h is Planck's constant). According to this relation the level width is indirectly proportional to the lifetime of the compound nucleus.

Since

$$h/2\pi = 0.65 \times 10^{-15},$$

then:

$$\Gamma = \frac{0.65 \times 10^{-15}}{\tau}. \tag{2.51}$$

For instance, the average time for the process (n, γ) in the case of slow neutrons is of the order of 10^{-14} to 10^{-15} sec. Therefore,

$$\Gamma = \frac{0.65 \times 10^{-15}}{10^{-15}} = 0.65 \text{ eV},$$

i.e., the level width for radiation of gamma quanta is of the order of a tenth of eV. In the case of high excitation energies when the captured particle possesses a high kinetic energy, the level width increases and the lifetime of the compound nucleus is shortened.

2.6.2 Cross-Section of Nuclear Reactions

Nuclear reactions take place when atomic nuclei of the target substance are bombarded with elementary particles. To express the yield we use the data of the *probability of an effective hit*. In characterizing this probability, we take as our basis a geometric representation of the bombarding particles as being points and the target nuclei as circular discs placed perpendicularly across the path of the incident particles. If we assume that every "hit" on the target leads to a corresponding nuclear reaction, its area will express the probability, and hence also the yield, of the reaction.

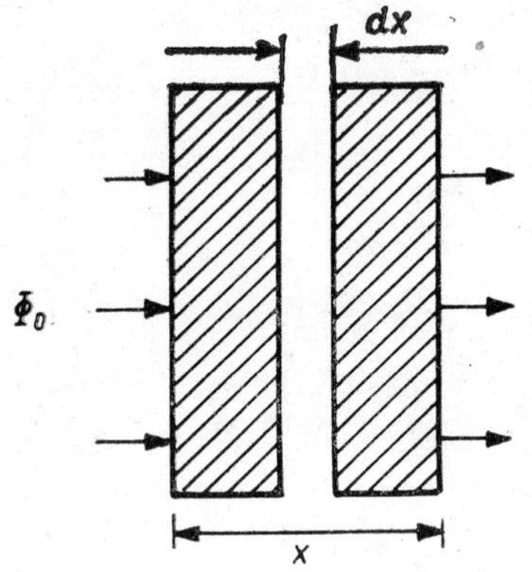

Fig. 2.36. Explanation of the cross-section of a nucleus.

The size of this area is called the *cross-section* (σ) *of the atomic nucleus* for the given nuclear reaction. According to this the cross-section refers to a particular nucleus and consequently it is defined as a *microscopic cross-section.*

Let us assume a uniform beam of bombarding particles hitting perpendicularly a unit area of the target substance under investigation. In the target of thickness dx, there are $N\,\mathrm{d}x$ atoms where N is the number of atoms per 1 cm^3 of target material (Fig. 2.36).

If σ is the microscopic cross-section of the nucleus for a certain nuclear reaction in cm^2 of effective area, the total effective area of the cross-section will be $\sigma N\,\mathrm{d}x$. Since the area of the target is 1 cm^2, the expression $\sigma N\,\mathrm{d}x$ indicates the reaction probability for any one incident particle with the atomic nuclei of the target, which has a thickness dx. We then have

$$-\mathrm{d}\Phi = \Phi\sigma N\,\mathrm{d}x, \tag{2.52}$$

$$\int_{\Phi_0}^{\Phi} \frac{-\mathrm{d}\Phi}{\Phi} = \sigma N \int_0^x \mathrm{d}x, \tag{2.53}$$

$$\ln \frac{\Phi_0}{\Phi} = \sigma N x, \tag{2.54}$$

$$\Phi = \Phi_0 \,.\, \mathrm{e}^{-\sigma N x}, \tag{2.55}$$

where Φ_0 is the incident flux, and Φ the flux of the particle beam after traversing a target of thickness x.

If the flux decrease is due solely to the effective capture of the bombarding particle within the nucleus, then

$$\Phi_0 - \Phi = \Phi_0(1 - e^{-\sigma N x}). \tag{2.56}$$

If x is very small, i.e., if thickness is infinitesimal, we obtain

$$\Phi_0 - \Phi = \Phi_0 \, \sigma N x. \tag{2.57}$$

The same simple expression may be arrived at also for a target of a finite thickness, if

$$\frac{\Phi_0 - \Phi}{\Phi} \ll 1 \tag{2.58}$$

so that the number of effective "hits" is so small that the flux of bombarding particles is but negligibly reduced in its passage through the target substance.

The cross-section has the dimension of the target area. As unit of measurement we use the value of 10^{-24} cm^2, called a *barn* (b). One *millibarn* (mb) is 10^{-3} and one *microbarn* (μb) is 10^{-6} barn.

The value of the cross-section does not depend solely upon the geometric dimensions of the target atomic nucleus, but rather, and essentially so, on the properties and the structure of the target nucleus as well as on the type and energy of the bombarding particles. Hence, the size of the cross-section is not a universal constant characteristic of one type of atomic nucleus, for its value depends also on the type of the nuclear reaction. Consequently, certain cross-sections possess higher values than the corresponding geometric cross-sections of target nuclei and, vice versa, they may be substantially smaller than the geometric cross-sections. Cross-sections are in the range of 10^{-8} to 10^4 barns. The most common values are those between 10^{-4} and 1 barn.

In the case of charged elementary particles the cross-section is smaller than the geometric section of the nucleus. Alpha particles with a large charge have an extremely small cross-section – 1/100 to 1/1000 of the geometric cross-section. In the case of charged particles the cross-section augments exponentially with increasing energy, until it attains the geometric cross-section, but cannot increase beyond it.

A different situation obtains in the case of neutrons. The lower their energy and hence also their velocity, the longer will they stay in the vicinity of the nucleus and consequently, the greater is the probability of a nuclear reaction. In the energy range of the order of several tenths eV, the correlation between the cross-section of neutron reactions and neutron velocity (v_n) is given by a descending curve which is characterized for every case, by the simple relation

$$\sigma = k \frac{1}{v_n} \tag{2.59}$$

that is, the reaction yield is inversely proportional to the velocity or the square root of neutron energy.

In the bombardment of a monoisotopic target by nuclear particles there occurs, for the most part, not one single type of nuclear reaction, but several processes simultaneously. If the target substance consists of several types of isotopes, then the number of possible variants of nuclear reactions is substantially higher.

Microscopic cross-sections may be *total*, signifying the probability that an arbitrary nuclear reaction takes place, and *partial*, referring to individual partial reactions. Therefore, the total cross-section is made up of cross-sections for individual processes which then express the share of these individual processes in the total probability of interaction between the particle and the nucleus. Most frequently the total cross-section (σ_t) is made up of the cross-section for scattering σ_s (the bombarding particle remains identical before and after interaction) and that for reaction σ_r. Hence,

$$\sigma_t = \sigma_s + \sigma_r . \tag{2.60}$$

The cross-section of a given isotope of a certain element is designated as the *isotope cross-section*: if it refers to the natural isotopic mixture, then we speak of it as the *atomic cross-section.* If the element comprises n isotopes whose cross-sections for a definite type of processes are $\sigma_1, \sigma_2, \ldots, \sigma_n$, then the atomic cross-section is given by the expression

$$\sigma_{at} = \sum_{i=1}^{n} \frac{p_i(\sigma_{iz})_i}{100}, \tag{2.61}$$

where p_i is the percentual representation of individual isotopes.

Frequently, the so-called *macroscopic cross-section* $\sum$ is employed, which is given by the product σN, where N gives the number of nuclei per 1 cm^3.

2.6.3 Resonance Effects

Among the most remarkable results of studies on nuclear reactions belongs undoubtedly the finding that the cross-section significantly increases at certain energies of bombarding particles (protons, deuterons, neutrons and others). This implies that for certain energies the capture probability of a particle by a nucleus, which thus leads to the formation of a compound nucleus, is unusually high. Here we speak of the so-called *resonance process of nuclear reactions.* This resonance effect is particularly manifest in nuclear reactions involving slow neutrons, as will be seen presently.

A theoretical explanation of resonance effects rests on the assumption that resonance occurs when the energy of the incident particle is such that the excited state of the nucleus, which ensues following particle capture, has an energy close to that of some quantum level in the nucleus.

The resonance process is schematically illustrated in Fig. 2.37 [1]. The system of

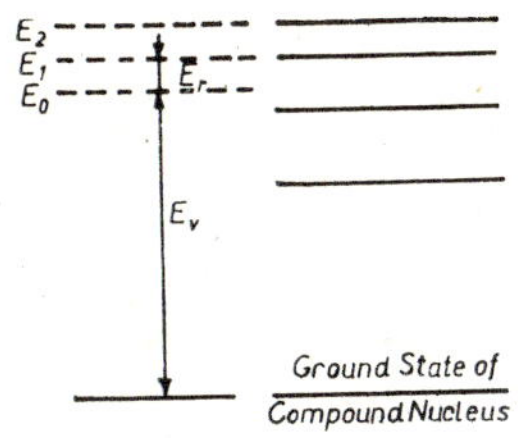

Fig. 2.37. The resonance process.

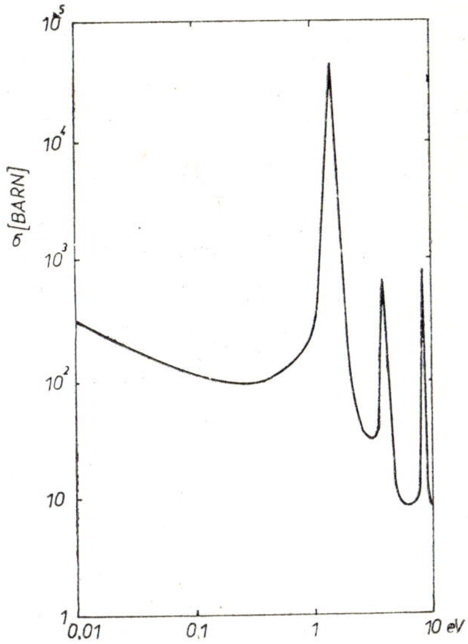

Fig. 2.38. Total cross-section of indium.

horizontal lines in the right-hand section of the figure represents the energy levels of the compound nucleus. Line E_0 on the left represents the energy of the compound nucleus formed following absorption of a neutron with zero kinetic energy. In this case the excitation energy of the nucleus is given only by the binding energy of neutrons in the compound nucleus. It is evident from the figure that energy E_0 in this case does not correspond to any quantum level of the compound nucleus. If, however, the neutron possesses a certain kinetic energy which together with its binding energy will suffice to excite the compound nucleus to energy E_1, then the energy of the excited nucleus will agree with that of one of its quantum states and the probability of a nuclear process ensuing will be very high. Therefore, resonance will occur if the neutron has a kinetic energy $E_r = E_1 - E_0$. Similarly, resonance will take place when the neutron kinetic energy is $E_r = E_2 - E_0$.

An example of a nuclear reaction with resonance maxima is given in Fig. 2.38. This is a correlation between the cross-section of indium and neutron energy. The course of this reaction is characterized by resonance peaks at 1.44 eV and two additional resonance peaks at higher neutron energies. Excluding the resonance range the cross-section correlation obeys the law "$1/v_n$."

The resonance process has been quantitatively dealt with by Kurtchatov et al. [13], while the theory of absorption resonance of slow neutrons has been worked out by Breit and Wigner [14]. They set up the relation which correlates cross-sections with the width of energy levels, and suitably expresses the correlation between cross-section and energy in the sphere of resonance

$$\sigma = \frac{\lambda^2}{4\pi} \frac{\Gamma_n \Gamma_\gamma}{(E - E_r)^2 + \frac{1}{4}\Gamma^2}, \tag{2.62}$$

where Γ_n and Γ_γ are partial widths of level for compound nucleus decay with neutron or photon emission, Γ is the total level width as given by the sum of partial widths $\Gamma = \sum \Gamma_i$, E the total energy of the incident particle (neutron) – inner and kinetic, E_r the energy value at which resonance is produced with the quantum level of compound nucleus, and λ the wavelength of the incident particle (as a neutron is involved, $\lambda^2 = k/E$, where k is a constant).

The Breit-Wigner relation shows that resonance maxima corresponding to high cross-sections occur solely in low energy ranges. The width of the resonance maximum depends upon the lifetime of the excited compound nucleus. If the lifetime is long, the maximum is narrow with a large cross-section, and conversely, if the nucleus is short-lived, the resonance maximum is wide and the cross-section small. When neutron energies are greater than the resonance energy, total cross-sections are very small (at 1 MeV they have a value of a few barns).

2.6.4 Types of Nuclear Reactions

Nuclear reactions may be classified according to several aspects. We shall mention a few of them:

1. According to *particle energy*, which induces reactions, we speak of nuclear reactions as low energy (of the order of MeV), medium energy (of the order of 100 MeV), and high energy (above 1 GeV). *Simple nuclear reactions* are generally realized by nuclear bombardments with nuclear particles of a relatively low energy (< 10 MeV). With increasing energy of the bombarding particle, nuclear processes become more complex; there is a growing number of emitted particles and these may be of various types. Up to now two types of complex nuclear reactions are known, *spallation* and *fission reactions*.

2. According to the *type of particles* inducing the reaction – e.g., reaction of char-

ged particles, gamma quanta, neutrons. The principal types of reactions according to the nature of the bombarding particle are given in Table 2.3.

Table 2.3

Main Types of Nuclear Reactions According to the Type of Bombarding Particles

Bombarding particle	Products and types of reaction						
	α	p	n	$2n$	γ	f	s
n	n, α	n, p	—	$n, 2n$	n, γ	n, f	n, s
p	p, α	—	p, n	—	p, γ	—	—
d	d, α	d, p	d, n	—	—	d, f	d, s
α	—	α, p	α, n	$\alpha, 2n$	—	α, f	α, s
γ	γ, α	γ, p	γ, n	$\gamma, 2n$	—	—	—

3. According to *the type of the nuclei* on which the reaction takes place—e.g., reaction on light ($A < 25$), on medium ($25 < A < 80$) and on heavy nuclei ($A > 80$).

In the following section we shall deal with some types of nuclear reaction.

Ordinary nuclear reactions belong among the most significant types of nuclear processes. The general schema of an ordinary reaction is of the form $X(a, b)Y$. In these reactions the target and the resulting nuclei may differ as to the atomic or the mass number, or even as to both. Hence, these reactions produce resultant nuclides of a different element from the target nucleus, eventually isotopes of the same element as the target nucleus. The bombarding (a) and the emitted (b) particles differ both in their nature and their number. Ordinary reactions may be connected by gamma quantum emission if the resultant nucleus is in an excited state.

Capture reactions or *radiation capture* are nuclear processes in which the target nucleus absorbs the bombarding neutron to form a compound nucleus which emits excess energy in the form of gamma quanta. The most frequent type of radiation capture is the reaction:

$$^{A}X(n, \gamma)^{A+1}X.$$

A schema of the mechanism of a capture reaction (n, γ) is given in Figure 2.39.

Nuclear fission reaction is brought about by a neutron or some other particle. Reactions involving atomic nuclear fission have a special mechanism. The liquid drop model of the nucleus envisages the target nucleus as oscillating following absorption of the bombarding particle. The original nearly spherical shape of the target nucleus changes, under the effect of oscillation, into a rhythmic shape of an elongated spheroid (see Fig. 2.40). Under sufficiently large oscillations the nucleus may break up into two fragments: $X(n, xn)\,Y_1, Y_2$. Fission reactions are generally indicated by the symbol (a, f).

In this reaction, then, the nucleus is broken up into two parts which have essentially different properties from those of the parent nucleus. If the energy of the incident particle is high, it increases probability of a *symmetrical fission* of the atomic nucleus in which two nuclei with approximately even masses are produced. If, on the other hand, the energy of the incident particle is low (e.g., thermal neutron), *an asymmetric fission* of the atomic nucleus occurs, in which the resulting nuclei have a different mass. In the fission of heavy nuclei, considerable energy is released and 2 to 3 fast neutrons are liberated for every nucleus that is split (e.g., for uranium isotopes ^{235}U and ^{233}U this amounts to 2.46 and 2.61, respectively, and in the case of plutonium

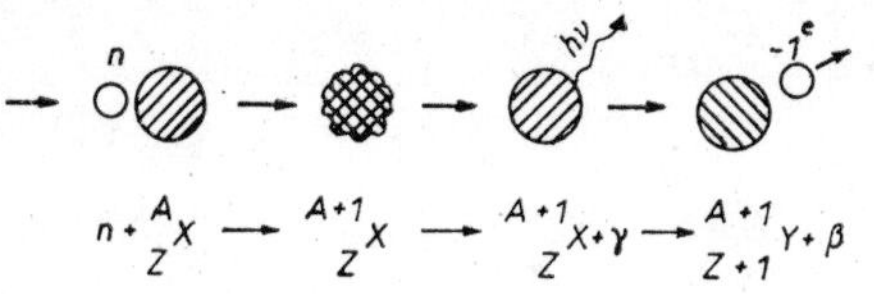

Fig. 2.39. Schematic representation of the reaction mechanism in radiation capture (n, γ) in four stages (*a*—neutron absorption, *b*—compound nucleus formation, *c* — transition to ground state through gamma quanta emission, *d* — emission of beta particle with formation of stable isobar). Stages *a*, *b* and *c* follow in close succession.

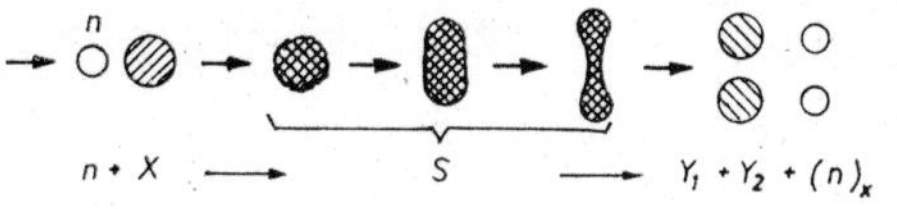

Fig. 2.40. Schematic representation of the mechanism of nuclear fission (*S* — compound nucleus).

isotopes ^{239}Pu and ^{241}Pu, it is 2.9 and 3.1, respectively, of fast neutrons per fission). These newly produced neutrons enable chain reactions to take place.

A *chain reaction* is a fission of atomic nuclei in which the neutron flux in the system remains stable or augments with time without any outside intervention whatsoever. Under ideal conditions a uranium chain reaction would run according to the schema (given two neutrons in the first generations) as shown below

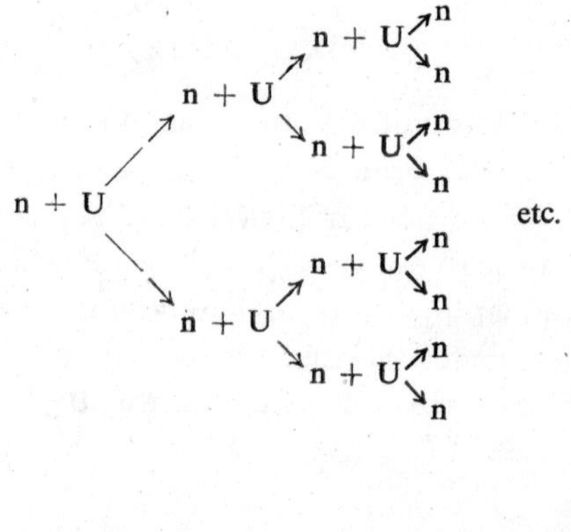

In spite of the fact that some of newly formed fast neutrons per absorption are absorbed in fissile isotopes, but do not enter into fission, and also that some are absorbed in non-fissionable substances, their number can be maintained sufficiently higher than 1, in order to compensate both for neutron escape from the system and also for the resonance absorption in deceleration and to enable a controlled chain reaction to be realized. Controlled chain reaction is used for producing energy in nuclear reactors (see p. 53).

Spallations occur when the target nucleus is bombarded with high energy particles (> 100 MeV). The target nucleus emits a large number of nucleons in the process. Hence the resulting nucleus differs considerably from the target nucleus in atomic and mass numbers. Spallations embody various partial processes which give rise to a great variety of products. Generally they can be designated as $X(a, xp + yn)\,Y$. The bombarding particle (a) may be alpha particle, proton, deuteron, gamma quantum and others; x is the number of emitted protons, and y that of neutrons. This schema includes only that part of splitting reactions in which only protons or neutrons are released. In general, spallation reactions are indicated by the symbol (a, s).

In *thermonuclear reactions* a condensation of light nuclei takes place with production of the resultant heavy nucleus and eventual emission of a particle. Schematically these reactions can be expressed by the equation

$$X_1 + X_2 \rightarrow Y(+b).$$

References

1. Drška L., Klimeš B., Slavík J. B., *Základy atomové fysiky (Principles of Atomic Physics)* Nakl. ČSAV, Prague 1958.
2. Vintr Z., *Přeměny atomového jádra (Nuclear Atom Conversions)*. SNTL, Prague 1966.
3. Goldansky V. J., Leykin E. M., *Přeměny atomových jader (Conversions of Nuclear Atoms)*. SNTL, Prague 1961.
4. Reactor Experiments Inc., Bulletin VF-5, Feb. 15, 1963.
5. Majer V., *Základy jaderné chemie (Principles of Nuclear Chemistry)*. SNTL, Prague 1961.
6. Buchanan J. D., Atompraxis **8,** 272 (1962).
7. Buchanan J. D., Activation Analysis with a TRIGA Reactor, General Atomic 2662 (1961). Intern. Conf. Modern Trends Activation Anal., College Station, Tex. 1961.
8. TRIGA, General Dynamics, General Atomic Division, Internal Purposes Series, August 1964.
9. Dunning K. et al., Phys. Rev. **48,** 704 (1935).
10. Vlasov N. A., *Neytrony (Neutrons)*. Gos. izd. techn. teoret. lit., Moscow 1955.
11. Bohr N., Nature **137,** 344 (1936).
12. Breit G., Wigner E., Phys. Rev. **49,** 519 (1936).
13. Kurtchatov I., Artsimovich L., Zh. exp. teor. fiz. **5,** 449 (1935).
14. Breit G., Wigner E., Phys. Rev. **48,** 918 (1935).

3. Introduction to the Nuclear Radiation Measurement Technique

V. KRIVÁŇ, J. TÖLGYESSY

Analytical methods, making use of nuclear radiation, yield information on the composition of substances. Measurement of nuclear radiation involves a variety of methods. All of them utilize some of the interaction of nuclear radiation with the detector material. The most significant methods utilize the ionization, scintillation and photographic effects. Of the various types of nuclear radiation the one most frequently met with in measurements is radioactive radiation.

Hence, we shall take up the most important methods that are currently applied in work with radioactive material.

This chapter cannot give a full account of the measurement techniques and their present state. For the details we refer the reader to other books and monographs [1–15]. We shall therefore discuss here only the principles of measuring techniques, the properties of the more important detectors and data-processing instruments to an extent adequate to give the analytical chemist a general knowledge of the subject, including practical advice.

3.1 Detection Techniques

3.1.1 Ionization Techniques

These methods are based on the ionizing effect of radioactive radiation. The most significant among them make use of ionizing effects in gaseous media. Consequently, we shall consider here only ionization in gases.

It is important first to characterize the discharge processes within the whole range of voltage in which ionization methods work.

When radioactive radiation spreads through a gaseous medium it causes part of the gas molecules to become ionized. The mechanism of ion formation differs according to the type of radiation and its specific interaction with the gas. Alpha and beta particles ionize directly along their course. These, being charged particles, either attract or repel the rather mobile electrons of the electron shell of a neutral atom or molecule. The action potential of these charged particles may be so powerful that one or more electrons may leave the atom or molecule and positive ions are formed. The ionizing particle thus loses a certain part of its energy, which is equal to the sum of the ionizing energy of the particular atom and the kinetic energy of the split-off electron. The longer the particle remains in the vicinity of the molecule, the greater will be the probability of its ionization; hence, the ionization is indirectly proportional to the velocity of the moving charged particle. The higher the charge of the particle, the greater the effect it will exert on electrons of neutral atoms and thereby the greater will be the probability of ionization. That is why alpha particles possess much stronger ionizing effects than beta particles. Free electrons formed through ionization may become bound to atoms and molecules, giving rise to negative ions.

Gamma radiation does not ionize directly but through photoelectric effect, Compton scattering, or pair production; the secondary electrons being formed give rise to ionization, as beta particles do.

When a gas exposed to a radiation is between two charged electrodes, the ions produced are affected by the electric field. The ions move towards the electrodes, and transmit to them their charge, and an electric current flows through the whole circuit. This current is proportional to the number of ion pairs reaching the electrodes and is called the ionization current. The radiation source and the experimental set-up being constant (the same shape, size and distance of electrodes) the number of ion pairs (n_p) reaching the electrode depends on the applied voltage (U). If the ionization effects of radioactive radiation are to be utilized, the relation $n_p = f(U)$ must be known. An examination of Fig. 3.1 shows that the curve can be divided into 4 regions:

Region *I* is usually termed the region of Ohm's law. Here, the ion velocity due to the effects of the electric field is so small that it can be neglected in view of the random movement of the molecules. The number of ions combining in their path with ions of opposite charge to form neutral atoms or molecules (recombination) considerably exceeds that of ions transmitted to the electrodes, which is directly proportional to voltage. Region *II* is the range of the saturation current. Ion velocity in this region, due to the effects of the electric field, is many times higher than that of the random movement of molecules. Practically all the ions are attracted by the electrodes before they can recombine. Hence the number of ions reaching the electrodes does not depend on the voltage, but solely on the intensity of radioactive radiation (activity of the sample). All the types of ionization chambers work in this region. A further increase in voltage brings us into a region where the so-called gas amplification is applied. Ions produced by the effects of radioactive ra-

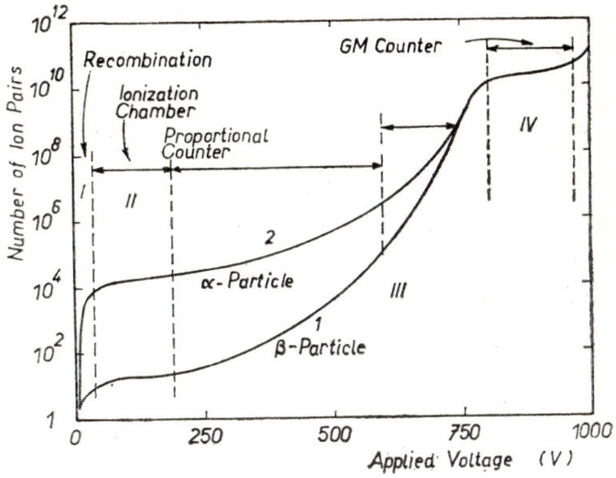

Fig. 3.1. Number of ion pairs incident upon the electrodes, as a function of applied voltage.

diation gain, on their path towards the electrode with an opposite charge, a higher kinetic energy than the ionizing energy of the given gas, and they themselves ionize the gas, just as the original particles. These newly created electrons are likewise accelerated, collide with gas molecules and ionize further. This brings about a collision ionization, even an avalanche or cascade effect, which is more powerful, the greater the intensity of the electric field. The number of primarily produced ions is thus increased many times. Initially, the total number of produced ions remains, even when increased through a collision ionization, directly proportional to the number of primarily produced ions. For this reason region *III* is designated as the proportional region in which proportional counters operate. With increasing voltage, the proportionality between the total number of ions and that of the primarily produced ions gradually ceases until finally the so-called Geiger-Müller region *IV* is reached in which the total number of ions does not depend at all on the number of primarily produced ions. Primary ionization is negligible and without significance in comparison with the powerful secondary ionization. A single primarily produced electron gives rise to approximately the same total number of ions as does any other ionization. Region *IV* is the operative region of Geiger-Müller counters.

3.1.1.1 Ionization Chambers

An ionization chamber consists of a gas-filled vessel in which two oppositely charged metal plates are fixed, to act as electrodes. The battery voltage is adjusted so that the conditions correspond approximately to the middle of region *II* (Fig. 3.1). The total current is then practically independent of voltage but depends solely on the intensity of radiation. Ionization chamber instruments fall into two categories:

With higher radiation intensities (sample activity), we measure the total ionization current passing through the chamber by means of a very sensitive galvanometer, electrometer, or an electronic amplification device. These chambers are called ionization chambers with constant ionization.

When low activities are involved, instead of measuring the current we may count individual current pulses produced by the entrance of a charged particle into the gas-filled chamber. This is possible, however, only when the current pulses have a certain amplitude. In practice this method is used only for measurements of alpha radioactive samples, because alpha particles have a high value of specific ionization. The chambers utilized in this method are called pulse ionization chambers, or also counting ionization chambers.

Construction of Ionization Chambers

Ionization chambers are designed in various shapes and sizes according to the type of radioactive radiation and the aim they are to serve, as well as according to the type and pressure of the gas filling. As filling material we may use air, argon and other gases at various pressures—greater or lees than 1 atm. The most common operating gas is air at normal pressure. All ionization chambers are made of metals and two electrodes which are insulated from the walls. Generally, the outer metal wall forms one of the electrodes and the other is inserted inside. Some ionization chambers have the shape of a board condenser with the shell forming one of the electrodes, or both are inserted within the chamber (Fig. 3.2a, 3.2b). In another type of these chambers the cathode is the outer, usually cylindrical shell and the anode is a metal rod in the longitudinal axis of the cathode (Fig. 3.2c). The anode should have a very smooth surface. Other types beside these are also in use, for example, a semispherical ionization chamber (Fig. 3.2d) etc. All these types of ionization chambers are suitable for measuring alpha and beta radioactive materials.

The current flows in the ionization chambers are of a very low intensity and this requires extreme care in the insulation between the electrodes. Insulators currently employed in electrotechnique are inadequate. The most commonly used materials are amber and sulphur. A drawback in the latter case is its high sensitivity to light. Recently these insulators have been replaced by organic plastic materials which proved to have outstanding properties in this respect. Frequently, a metal ring is placed within the insulators and is earth-connected.

In view of the small range of alpha particles in gases, ionization chambers for

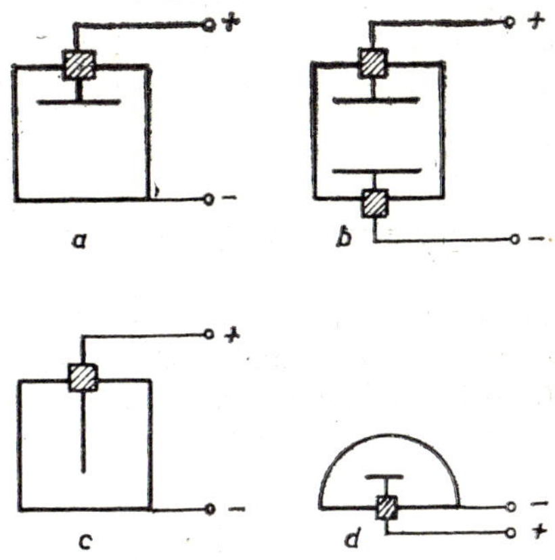

Fig. 3.2. Some types of ionization chambers (*a*, *b*, — flat, *c* — cylindrical, *d* — semispherical).

measuring alpha radioactive preparations are smaller. The dimensions of the ionization chamber are such that the range of the alpha particles comprises in its sensitive volume. The sample to be measured is always placed directly into the chamber.

Chambers intended for measuring beta radioactive samples are substantially larger. The samples may be placed either inside or outside the gas-filled chamber and beta radiation penetrates its effective operation space through a window of a thin aluminum layer or other material.

To measure gamma radiation from radioactive samples we generally make use of ionization chambers whose wall, forming the cathode, is of a material with a high atomic number and of a thickness corresponding to the range of electrons produced by gamma radiation. Electrons discharged from the cathode then ionize the gas-filled chamber. For this reason, gamma radioactive samples are generally placed outside the ionization chamber during measurement.

Measurement of Ionization Current

Ionization chambers in which the gas becomes ionized under the effects of radioactive radiation, must be connected with a suitable measuring device that enables the intensity of the saturation current to be recorded. In every measuring system, we must, in the first place, determine the volt-amper characteristic of the ionization chamber, that is, the correlation between the current I and voltage U. This volt-amper characteristic then allows us to determine the working voltage of the ionization chamber in the region of the saturation current. The saturation ionization current is generally very weak, of the order of 10^{-15} to 10^{-8} amperes, and therefore very sensitive measuring devices are required.

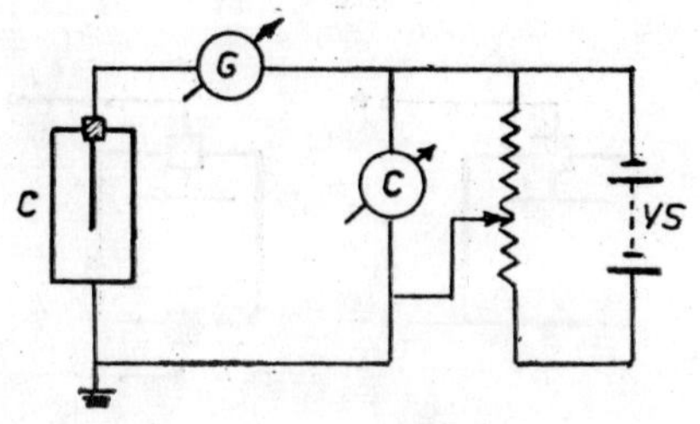

Fig. 3.3. Direct measurement of ionization current by a very sensitive galvanometer (*IC* — ionization chamber, *VS* — voltage source, *C* — voltmeter, *G* — galvanometer).

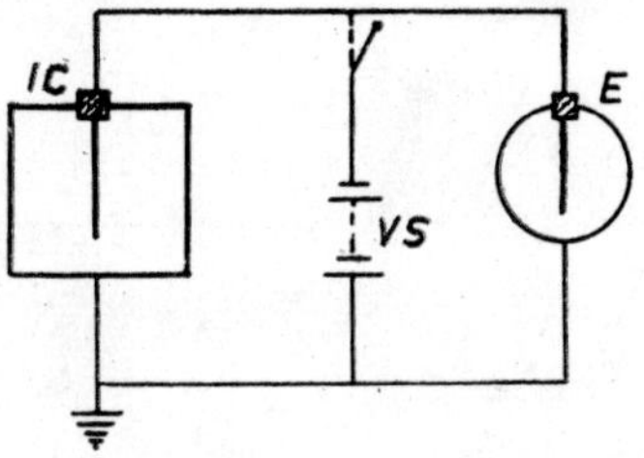

Fig. 3.4. Measurement of ionization current by an electrometer (*IC* — ionization chamber, *VS* — voltage source, *E* — electrometer).

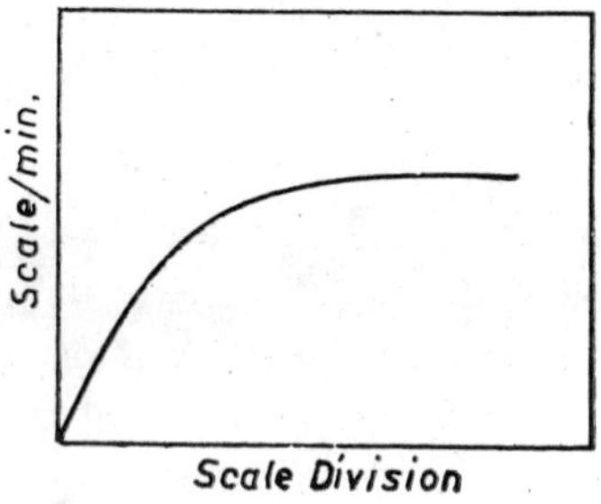

Fig. 3.5. Characteristics of an electrometer.

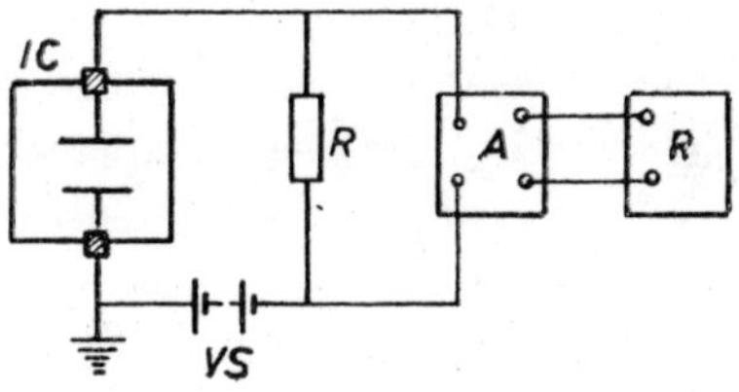

Fig. 3.6. Pulse ionization chamber connection (*IC* — ionization chamber, *R* — resistance, *A* — amplifier, *R* — recording device, *VS* — voltage source).

In the case of samples of higher activity, the ionization current flowing between the electrodes of the ionization chamber may be measured directly by means of a very sensitive galvanometer connected to the circuit (Fig. 3.3). However, the ionization current is generally measured following its preliminary amplification by means of some electronic reinforcing device.

With lower radiation intensities, we make use most frequently of electrometers. Any type of electrometer may be employed to this end, for instance, the classical leaf electrometer, Wulf's one- or two-fiber electrometer, Laurithen's torsion (quartz-fiber) electroscope, and others. The anode of the ionization chamber is connected with the fiber system of the electrometer (Fig. 3.4), and instrument is charged either from an a.c. voltage source with a rectifier, or from a battery, to a certain potential. The electric field causes the fiber in the electrometer to be deflected, the range of this deflection depending on the magnitude of the charge. In measuring a sample, radioactive radiation ionizes the gas-filled chamber, which becomes electrically conductive. The whole system thus becomes discharged and the fiber returns along the scale to its initial position. From the rate of movement of the fiber, we may determine the value of the ionization current, or the activity of the sample. When electrometers are used for the measurements, we must know their characteristics: the correlation between the rate of movement in divisions per unit of time, and scale divisions $d/t = f(d)$. The operational region on the scale is represented by the straight line sector on the graph (Fig. 3.5). In absolute measurements, calibration to volt sensitivity must be made and the value of electrometer capacity determined.

In the case of very weak ionization currents, two equal ionization chambers in compensated coupling are employed. Such measurements give a high degree of sensitivity.

Among the measuring techniques used with ionization chambers is that of constant deflection. Here, an electrometer measures the voltage drop produced on a high-value resistance ($10^{11} - 10^{13}\,\Omega$) connected in series with the ionization chamber circuit. The deflection in the electrometer is directly proportional to the current. An electrometer may be replaced by a high-sensitivity electronic voltmeter.

Pulse Ionization Chambers

It has been stated that pulse ionization chambers are used to measure very low activities. We describe here the function of a pulse ionization chamber with reference to the scheme of coupling in Fig. 3.6. When a charged particle enters the ionization chamber, ionization immediately reacts to a maximum value and then declines to zero. The time change in ionization causes a time change in the charge on the electrodes as well as a time-variable voltage. This becomes manifest as a time-variable current in the circuit which evokes a voltage pulse on resistance R. Voltage pulses become amplified in a linear amplifier A and are recorded by a suitable recording device R. However, this can be accomplished only when the pulses have a certain minimum size – in practical application, mainly in the measurement of alpha radiation, which is noted for its high ionizing capacity. This technique is unsuitable for measuring beta and gamma radiations. As the intensity of current pulses originating in the chamber is directly proportional to the energy of alpha particles, the pulse ionization chamber can be used also for alpha spectrometry provided a suitable electronic equipment (pulse discriminator and pulse analyzer) is available. These chambers, in contrast to the preceding types, are filled with a gas that offers the least possible probability of negative ions being formed through electron capture by neutral molecules. The greater electron mobility in gas, as against negative ions, causes a faster pulse increase in the outer current circuit. In this way, pulses lasting a few microseconds are produced, so that the detector is characteristic for its good resolving power. However, when pulse ionization chambers are applied to spectrometric measurements, it should be borne in mind that pulse amplitude does not depend uniquely on the size of the charge produced, but likewise on the orientation and place of entry of the particle into the chamber. This inconvenience is remedied by interposing a grid between the electrodes, and charging it to a positive potential. Thus a 1 % energy resolving power can be achieved.

3.1.1.2 Geiger-Müller Counters

Geiger-Müller (GM) counters belong among the most significant nuclear radiation detectors. They remain unsurpassed when there is need for simple measurements of rather low sample activities. In the processing and evaluation of signals from GM counters, simple, low-cost and operationally reliable measuring equipment may be used, for these detectors do not require any high-stability feeding voltage and yield considerable outgoing current pulses. They are relatively stable and sensitive in performance, versatile in construction and design, and may be used for recording various types of nuclear radiation.

Function of Geiger-Müller Counters

The operational activity of GM counters is based on a powerful secondary ionization of the gas filling due to the effect of the electric field between the electrodes. In order to obtain the required intensity of the electric field necessary for gas amplification at technically easily accessible values of the input voltage, an electrode arrangement is used which is characteristic in that one electrode (the cathode) has a cylindrical shape and the other (the anode) the shape of a thin wire suspended along the axis of the cylinder. In this set-up the intensity of the electric field grows rapidly from the inner wall of the cylinder towards the surface of the anode wire, where it acquires its peak value.

If a certain voltage U (in V) is applied to the above electrode system, the intensity of the electric field E (in V cm^{-1}), at a distance d from the anode wire, may be given by the following formula

$$E = \frac{U}{d \ln \frac{r_1}{r_2}}, \tag{3.1}$$

where r_1 is the cathode radius, and r_2 the radius of the anode wire.

The electric field intensity is, therefore, very high in the vicinity of the anode wire; for current values of U, r_1 and r_2 it is of the order of 10^4 V cm^{-1}, and for this reason the gas amplification process is, for the most part, effected here.

On entering the counter, a charged particle or a photon causes ionization of the

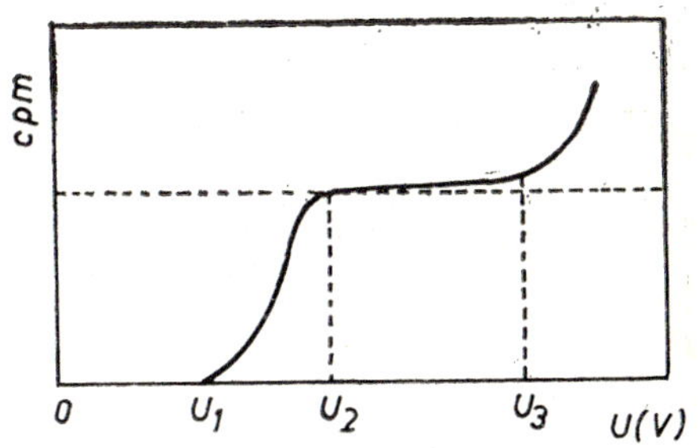

Fig. 3.7. Characteristics of a Geiger-Müller counter.

gas content. Under the effect of the electric field, the electrons produced move towards the anode wire and positive ions towards the cathode. The electric field imparts to the electrons on their short path close to the anode wire such high kinetic energy that they themselves cause further gas molecules to be ionized. The newly formed electrons acquire in their turn a high kinetic energy and cause further ionization; thus each electron will produce an avalanche. In an extremely short time interval (of the order of 10^{-8} sec from the start of the discharge) a whole cascade or "avalanche" of electrons reaches the anode wire of the counter. This represents an amplification of 10^8 to 10^{10} times the minute charge produced by the primary ionization. The current in the counter is passed through an external resistance and the voltage pulse is amplified and recorded with a suitable device.

The number of recorded pulses per time unit (counting rate, cpm) is a function of the voltage applied to the electrodes of the counter. This correlation is called the *characteristic of GM counter* (counting rate-voltage curve) and is illustrated in Fig. 3.7. The counter begins to detect incoming particles only when the threshold voltage has been reached. Initially, the counting rate rapidly rises with increasing potential. The amplitude of pulses obtained in the counter grows approximatively linearly with electrode potential, but in the GM region does so independently of the type and energy of detected particles. Counting devices record only pulses whose amplitude exceeds a certain threshold value. Amplitudes of pulses from the counter are not equal even at a constant potential, but fluctuate about a certain mean value. With a voltage below the starting potential, the amplitudes of all pulses are lower than the threshold value. In the plateau region ($U_2 - U_3$), practically all the pulses are higher than this value. The abrupt rise of the counting rate in the $U_1 - U_2$ region is due to the rapid growth of pulses with amplitude exceeding the threshold, with the potential. In the plateau region, the counting rate grows only moderately with increasing potential. The abrupt growth of the counting rate at the end of the plateau ($U > U_3$) may be ascribed to the pulses being produced in groups rather than individually, as may be seen from oscillographic recordings.

Following the ionization event or the discharge, the counter remains for a certain time insensitive and fails to react to the next particle that enters immediately after the discharge. The mobility of positive ions is approximately 1000 times smaller than that of electrons, as a result of which they do not succeed in disengaging themselves from the site of their production towards the cathode during the duration of the discharge. The wire is thus surrounded by a positive space-charge that lowers the intensity of the electric field in its vicinity. A new discharge may take place in the counter only when the positive ions have moved some distance from the anode and the intensity of the electric field regains the required value – a discharge leading to a new pulse with an amplitude exceeding the threshold value. The time interval required for a new discharge to take place is called the dead time of a Geiger-Müller counter. Its value is further affected also by the sensitivity of the counter. The lower the threshold amplitude value of the counter, the shorter is the dead time. It improves also when the mean amplitude value is increased by way of increased potential on the electrodes.

Owing to these factors, a certain number of particles fail to be recorded and this must be taken into account in the processing of the results. The higher the sample activity, the greater the proportion of unrecorded particles (photons). Therefore, the dead time correction must always be applied when higher activities are involved.

The following relation is valid for the total number of particles that pass through the counter per time unit:

$$n_0 = n + n \, . \, \tau \, . \, n_0, \tag{3.2}$$

where n is the number of particles recorded per time unit, and τ the dead time.

From this relation we have

$$n_0 = \frac{n}{1 - n \, . \, \tau} \, . \tag{3.3}$$

The dead time in the case of GM counters is usually of the order of 10^{-4} sec. This, however, depends also on the operational load placed on the counter.

Types of Counters According to their Gas Filling

In the process of neutralization of positive ions on the cathode, electrons may be released that provoke a cascade ionization and a discharge in the counter without a charged particle or a photon having entered from outside. These events occur when positive ions of simple molecules are discharged on the cathode, and may be prevented in various ways. The most frequently used devices are the so-called *self-quenching counters*. Various gases such as argon, helium, etc., are used in mixtures with a small quantity of an organic vapor (e.g., alcohols, ethers, ketones and others) under a pressure of about 100 mm Hg. The ionization potential of the polyatomic gas should be lower than that of the monoatomic gas. On their pathway to the cathode, ions of the monoatomic gas transfer their charges to the neutral polyatomic molecules, so that virtually only ions of the latter reach the walls of the GM tube. Polyatomic vapors have a further important property: they absorb ultraviolet radiations formed during electron transfer and reduce the number of incorrect pulses down to a minimum. During these processes, polyatomic molecules become excited. They get rid of their excitation energy, in particular by dissociation and decomposition, this is related to the finite lifetime of self-quenching counter tubes. Every discharge results in the dissociation of a large number of polyatomic molecules. A self-quenching counter tube of current construction becomes operationally ineffective following 10^8 to 10^{10} discharges. This is the principal drawback of these counters. The lifetime of a counter filled with organic quenching vapors may be prolonged when the lowest possible voltage is applied, since the size of the ensuing charge and thereby also the quantity of decomposed quenching molecules is less at a lower voltage.

A similar self-quenching effect can be observed when organic vapors are replaced by a halogen gas (chlorine, bromine, or iodine). These are known as *halogen counters*. The mechanism of the self-quenching effect is similar to that described above. Here, however, advantage is taken of the reversibility of dissociation of a diatomic molecule and hence the operational lifetime of the counter tube is not limited. Another advantage of halogen counters resides in their low operational voltage (about 300–400 V) and their suitability for work in a wide temperature range. On the other hand, they have a longer dead time and allow double pulses to occur, which makes them unsuitable when accurate measurements are required. They are mainly built into various portable instruments where the advantage of a low-voltage source can be utilized.

In some special cases, for instance when radioactivity is to be measured at very low or at high temperatures, *non-self-quenching Geiger-Müller counters* are employed. They are filled with simple mono- or diatomic gases, such as hydrogen, air, etc. The discharge is interrupted by a special electronic device which reduces the anode wire potential after 10^{-4} sec of each discharge.

The most important characteristics determining the quality of a GM counter are: operational characteristic, efficiency, background, resolving time and lifetime.

The *cpm-voltage characteristic* of a GM counter is obtained by plotting the counting rate against the voltage at constant radiation intensity (see Fig. 3.7). The interval $U_2 - U_3$ is called the plateau; this is the operational region of the counter. The gradient of the operational region is characterized as a percentage change in the counting rate per 100 V. Good-quality GM counters have a plateau length of at least 200 V and a gradient less than 5 %. The working voltage at a shorter plateau is selected in the middle of the plateau's length, as measured from the initial voltage; for a longer one, in the first third. The operational characteristic of a counter changes during its utilization. The operational region becomes shorter and its gradient increases until the counter ultimately loses its operational region entirely. This is related to alterations that take place in the gas filling of the self-quenching counter tubes. Therefore, the operational characteristic must be regularly checked and eventually the value of the operational voltage determined.

The *efficiency* of counters is defined as the ratio of the number of recorded particles or photons to that of all the particles (photons) that pass through the effective volume of the counter.

When beta particles are counted, the efficiency depends in particular on their energy, the ionization potential of the gas filling, and its pressure. The conditions under which GM counters operate are that beta particles reaching the effective volume create, with approximately 100 % probability, at least a few ion pairs, thus producing current pulses, and are recorded. Hence, the efficiency in counting beta particles is close to 100 %. Thus, for instance, efficiency of a counter filled with argon and alcohol vapors under a pressure of 100 mm Hg is, for any type of beta radiation, equal to 99.8 %. Efficiency for alpha radiation is also close to 100 %.

The efficiency in counting gamma quanta is very low. This is due to the fact that gamma quanta are counted by means of secondary electrons originating from the cathode and the filling of the counter. It may be somewhat improved by the use of a heavy cathode material. But even then it is of the order of 1 %.

The *background* of a GM counter is the average counting rate recorded at the operational voltage of the counter, without the presence of the radioactive sample. Sample activity is measured in special lead castles where the background count is given exclusively by cosmic radiation and radiation from natural radioactive isotopes present in minute quantities in the material of the counter and the castle. The background affects precision of measurement. This becomes markedly manifest in measurements of low activities. Therefore, we endeavor to obtain the lowest possible background count and introduce corresponding corrections in processing the results of measurement.

In measurements with GM counters we must know the region of temperature in which these counters can operate satisfactorily. Thus, for instance, one of the most wide-spread types of counters, filled with a mixture of argon and ethylalcohol, cannot be used at temperatures below 0 °C, because of the condensation of alcohol vapors. A change in temperature shifts the operational region. A temperature increase of 1 °C shifts the starting voltage by 3–6 V.

Resolving time is given by dead time and may be in the range of 10–300 μsec [2].

Construction of Geiger-Müller Counters

GM counters may be of various designs according to the type of radiation measured with them. Essentially they may be divided into two types, namely end-window and cylindrical counters (Fig. 3.8). End-window counters are used for measuring alpha and beta radiations. Such a counter consists of a glass tube closed with

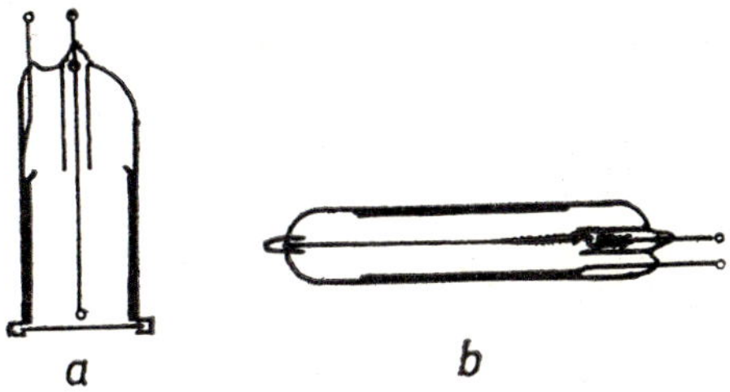

Fig. 3.8. GM counters (*a* — window, *b* — cylindrical).

a silica window, in which are located a cylindrical cathode of copper or brass and a thin-wire anode of tungsten, platinum or steel. The mica windows for alpha and soft beta radiations are very thin so as to allow particles to enter the inner space. The thinnest windows are of 0.5 mg cm^{-2} thickness, through which alpha radiation with a minimum energy of 0.6 MeV and beta radiation with the lowest possible energy of 0.02 MeV can penetrate. To count the high energy beta particles, cylindrical metal counters are used. The shell is prepared from a thin aluminum sheet, 0.1–0.15 mm in thickness. These counters are able to measure beta radiation from a maximum energy of about 0.5 MeV.

Gamma radiation is measured by means of cylindrical glass counters. These counters also measure beta radiation of a maximum energy higher than 1.7 MeV.

To measure very low energy beta radiation (and also alpha radiation) the samples (solid, liquid or gaseous) are inserted into the counter. Most frequently the counter is filled with the gaseous sample, or the sample is admixed to a suitable filling. The above technique yields the highest possible geometrical efficiency (4π).

To measure radioactive liquids we make use of special counters which may have a single jacket (during measurement they are immersed in a ring vessel with the solution), or a double jacket (the space around the counter formed by the two jackets is filled with the solution). The advantage of these double jacket counters is in their simplicity and outstanding reproducibility of the geometric conditions of measurement. Radioactivity of flowing liquids or of gases is measured with flow-counters. The radioactive solution or gas flows through a spiral around the counter, or the outer jacket of the tubes.

3.1.1.3 Proportional Counters

Proportional counters occupy a position midway between ionization chambers and Geiger-Müller counters, both in their functional principle and their construction and applicability. They operate in region *III* (see Fig. 3.1). They utilize the increased number of ions primarily produced by the secondary collision ionization (similar to GM counters); nevertheless, the size of the overall charge remains directly proportional to the number of primarily produced ions (similar to ionization chambers). As a result of the secondary cascade ionization the primarily evolved charge is amplified 10^2–10^7 times. This permits even weakly ionizing beta particles and gamma quanta to be detected and counted. The mechanism involved in amplification is entirely similar to that described for GM counters. In their construction too, the commonly employed closed proportional counters resemble Geiger-Müller counters (cylindrical and wire electrode system), the only essential difference being in the dimensions. Many gases cannot be used as filling material for proportional counters because of their electronegative character, involving the capture of electrons to form negative ions. The gases most frequently employed are methane, propane, and a mixture of argon and methane in the ratio of 1 : 1 to 9 : 1, as well as a helium–isobutane mixture.

However, in measuring low sample activities we usually give preference to Geiger-Müller counters and use ionization chambers when samples of high activity are counted. There are cases, nevertheless, where the use of proportional counters is more suitable, or the purpose of the measurement may be fulfilled only by proportional counters, from among all the ionization techniques.

Under certain conditions, a proportional counter is convenient for spectrometric beta and gamma measurements. Here one condition must be fulfilled: particles must discharge all their energy in the gas filling of the sensitive volume of the counter. With current dimensions and gas pressure, a proportional counter may be used spectrometrically to investigate beta particles with a maximum energy not exceeding 100 keV (under the most favorable conditions, 200 keV). Gas amplification must not be too high for it would give rise to a large number of positive ions forming the space charge which provokes saturation in the process of gas amplification. The proportionality between pulse amplitude and particle energy is thereby lost. The region of beta particle energy may be extended to 2 MeV in spectrometric measurements with proportional counters by taking advantage of the effect of axial

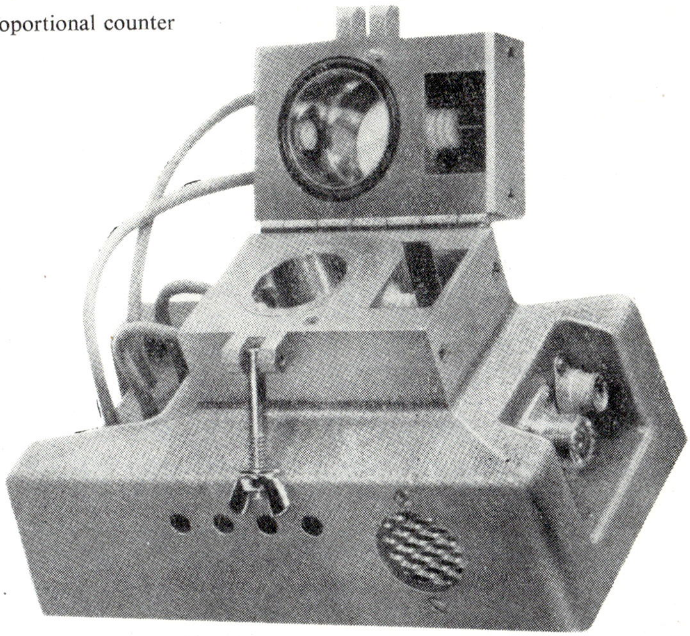

Fig. 3.9. Through-flow proportional counter (Labgear, Cambridge).

magnetic fields, which cause electrons to move along a spiral pathway [1]. The lower energy limit for beta spectrometry is essentially given by the amplifier noise, the properties of the gas filling, and the statistical fluctuation of the number of ion pairs produced by low-energy particles. The energy involved does not usually exceed 1 keV. In special cases, proportional counters are used in spectrometry of low-energy gamma and X-radiations on the basis of photoelectrons emitted under the effects of this electromagnetic radiation from the cathode walls into the sensitive gas volume of the counter.

Of great significance and wide applicability are *proportional flow counters.* They are not used in spectrometry, but only for activity measurements. Their essential advantage lies in that the radioactive samples are introduced into the counter and are measured directly, thus excluding absorption in intermediate layers (air and window of counter), and a high geometric efficiency may be attained, depending on their construction and the preparation of samples (2π or 4π sr). In this way even very soft beta radiation may be measured relatively conveniently. Flow counters are so designed as to be easily opened when the sample is being introduced. In order to exclude air and to maintain uniform discharge conditions, the counting gas is left to stream continuously through the counter before and during measurement (methane is the gas most frequently used) at atmospheric pressure. Proportional flow counters may have diverse geometric shapes, often even a noncoaxial arrangement without the cylindrical cathode. One of the designs commonly used is illustrated in Fig. 3.9.

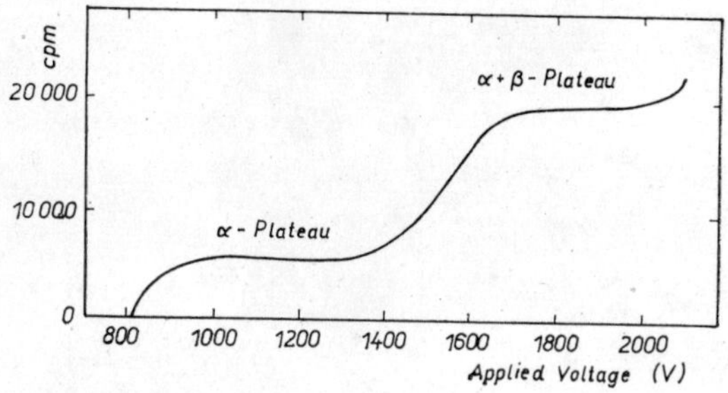

Fig. 3.10. Characteristic of a gas flow (methane) proportional counter.

Proportional counters may advantageously be utilized for measuring samples emitting alpha and beta particles. From the characteristic of the flow-through counter (for alpha and beta radiation) shown in Fig. 3.10, it may be seen that at low voltages only alpha particles are counted, because of their high ionization ability. Even a small amplification causes pulses to be produced with an amplitude exceeding the threshold value and an alpha plateau is formed. Higher voltage is needed to achieve gas amplification, at which beta particles provoke pulses with a sufficiently large amplitude for their measurement. Following a short rise the $\alpha + \beta$ plateau is reached. By means of two measurements at two different voltages (in the region of the first and the second plateau) first only alpha particles are counted and the alpha and beta particles together. In a similar way, we can measure, for instance, the activity of alpha radiation even in the presence of interfering beta and gamma radiation sources.

Flow-through proportional counters enabling detection in 4π geometry can be used for the measurements of absolute beta activity [1].

A general advantage of proportional counters is their good resolving time which is maximally in the order of 1 μsec. No correction for dead time is needed up to a counting rate of about 5×10^4 cpm. This is their great advantage over Geiger-Müller counters. A further priority is the low background which, for instance, in the case of alpha radiation detection can attain a value of the order of 0.1 cpm, which enables a marked improvement in the accuracy of measuring low activities. Detection efficiency for gamma radiation is as low as in GM counters (about 1 % resolution); for alpha and low-energy beta particles it is about 50 % when gas-flow counters are used [15].

3.1.2 Scintillation Methods

The scintillation method belongs among the earliest detection methods of radioactive radiation. The first instrument to work on this principle was Crookes' spinthariscope for counting alpha particles. The significance of scintillation methods rose enormously when the scintillation counters in their present-day form were introduced, i.e., when the flashes, viewed until then by the naked eye, could be observed by means of a sensitive photoselective device.

3.1.2.1 Working Principle of Scintillation Counters

A scintillation counter consists essentially of two basic parts: a scintillator and a photomultiplier tube. Their set-up is shown in Fig. 3.11.

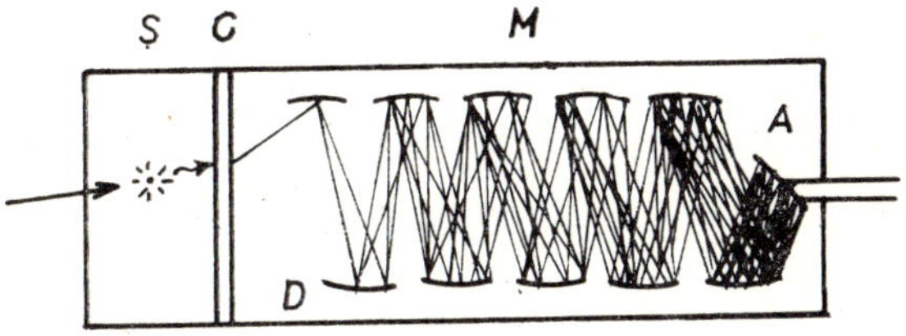

Fig. 3.11. Schematic representation of the activity of a scintillation counter (*S* — scintillator, *C* — photocathode, *M* — photomultiplier, *D* — dynodes, *A* — anode).

Charged particles, gamma and X-radiation, when passing through scintillator *S* lose their energy in the excitation and ionization of atoms and molecules. In their transition to lower energy levels and ultimately to the ground state they emit the absorbed energy in the form of photons with a wavelength in the visible band, or its neighborhood. At the same time the angular distribution of the photons is entirely irregular. Some of them reach the photocathode *C* of the photomultiplier *M* and liberate photoelectrons from it. The emission of photoelectrons from the photocathode is so weak that these could not be recorded directly and are therefore first multiplied in a photomultiplier tube. This consists of a group of electrodes (dynodes) *D* whose surface is sensitive to the secondary emission of electrons following the impinging primary electrons. Every successive dynode has a more positive potential than the preceding one. Electrons released from the photocathode are accelerated by a potential of 100 V or more to the first electrode where each one produces several (n) secondary electrons; these secondary electrons are similarly accelerated and increased n-fold at the second dynode, etc. The number of original electrodes is increased 10^6–10^8 times. Finally an intensive electron stream hits the anode of the photomultiplier tube A; in this manner, sufficiently large pulses are produced, which can be conveniently detected and recorded.

The function of a scintillator can be fulfilled by substances having luminescent properties. Two types of luminescence are known: *fluorescence,* in which emission occurs in a time interval smaller than 10^{-8} sec following primary excitation (that is the lifetime of an excited atom), and *phosphorescence*, in which the lifetime of excited atoms is greater than 10^{-8} sec (sometimes up to several minutes). Scintillation techniques makes use of scintillating substances that possess both fluorescent and phosphorescent properties.

A scintillation counter has three essential advantages over Geiger-Müller counters:

A. Its efficiency of detection for gamma radiation is higher because the probability of its absorption of the radiation penetrating into the scintillator is higher than that of ionization in the case of gas filling, or the production of photoelectrons and their emission from the wall of the counter.

B. Scintillation counters have a very short dead time. Hence, the resolving time of the whole counting equipment is determined by its electronic set-up. When two particles hit the scintillator simultaneously, the outgoing pulse equals the sum of the two pulses which each particle would produce separately.

C. Pulse amplitude is a function of energy, so that pulses may be utilized for spectrometric measurement. However, the instruments necessary for processing pulses are more sophisticated than in the case of GM counters.

3.1.2.2 Scintillators

Substances employed as scintillators must fulfill certain requirements, namely:

a) they must strongly absorb primary radiation,

b) the light yield must be as high as possible,

c) they must minimally absorb light produced by deexcitation and deionization,

d) the spectrum of this light must lie within the spectral sensitivity of the photomultiplier tubes and

e) the light flash should be as short as possible.

Scintillators may be divided into four groups: inorganic crystals, organic crystals, organic liquids and emulsion scintillators.

The luminiscent properties of inorganic scintillators are related to the existence of the crystal lattice. Inorganic crystals are used in particular for measuring the activity of gamma radiation sources and energy determination of gamma and X-radiation. All involve elements with a high atomic number thus assuring a high absorption efficiency through photoeffect. The flash duration (dieary time) averages 10^{-6} sec. Although this value is lower by an order of 3 than the average value in organic scintillators, it still represents a marked improvement in the resolving time in comparison with GM counters (by an order of 2). In general, inorganic scintillators applied in practice can be classified as follows: halogenides, sulfides and tungstates.

The most widely used inorganic scintillator is NaI, activated with about 1 % Tl. One drawback of this scintillator is that it is hygroscopic and therefore should be kept in an air-tight container. Sodium iodide gives a very high light yield, and has the shortest dieary time of all the inorganic scintillators. Another very important scintillator with good properties is CsI activated with Tl; it is not hygroscopic. Certain special inorganic scintillators are suitable for the detection of alpha radiation, e.g. ZnS(Ag), for neutron radiation, e.g. ZnS(Ag) – plastic substance, ^{6}Li – ZnS(Ag), LiI(Cu) and others.

The most important inorganic scintillators and their properties are summarized in Table 3.1. It should be noted, however, that the dieary time and the intensity of the emitted light depend on the temperature [3]. Unless otherwise stated, the values refer to room temperature.

While in inorganic scintillators luminescence is a property of inorganic crystals, in the case of organic scintillators it is conditioned by monomolecular processes. This makes it possible for organic scintillators to be prepared in the form of crystallic substances, plastic substances, solutions and gels. In comparison with inorganic crystals, organic scintillators are marked by a substantially shorter dieary time, which is, for the most part, on the order of 10^{-9} sec. This group of scintillators has further priorities. They generally have a high permeability to emitted light. The wavelength of fluorescent radiation lies in the sensitivity region of presently used photomultiplier tubes. Scintillators of arbitrary size may conveniently be produced from suitable compounds and plastic matter. Nevertheless, this group of scintillators also possesses serious drawbacks, in comparison with inorganic scintillators. In the first place, they are unsuitable to detect gamma radiations and measure their energies, since all these scintillators have a low specific weight (about 1 g cm^{-3}) and consist of atoms with a low atomic number, essentially of hydrogen and carbon, so that there is little probability of gamma radiation being absorbed. Another disadvantage is that the light yield of gamma and X-radiation is smaller than, for example, in NaI(Tl) and ZnS(Ag), and in the case of heavy charged particles, is energy dependent. For these reasons organic scintillators are best suited for measuring beta samples.

The most important among organic crystals is anthracene (it serves as standard of reference for the light yield of all other scintillators). Also naphthalene, stilbene, tetraphenyl and others are applied. They are included in Table 3.2.

Recently numerous so-called plastic scintillators have been developed which are

Table 3.1

Crystalline Inorganic Scintillators and their Physical Properties

Scintillator	Density	Melting point (°C)	Light yield (% in terms of anthracene)	Dieary time (sec)	Wavelength at maximum emission (Å)	Application
NaI (Tl)	3.67	650	230	0.23×10^{-6}	4130	γ, X
NaI (pure)	3.67	651	440	0.06×10^{-6}	3030	γ, X (rapid counting)
Cs (Tl)	4.51	620	95	0.7×10^{-6}	4200–5900	γ, charged particle
Cs (pure)	4.51	621		0.6×10^{-6}	4000	γ (low energy), charged particle
Li (Eu)	4.06	445	75	1.2×10^{-6}	4750	n
CsF (pure)	3.59	684	12	5×10^{-9}	3900	γ (rapid count)
ZnS (Ag)	4.09	1850	300	0.2×10^{-6}	4500	α
$CaWO_4$	6.1	1545	36	6×10^{-6}	4300	γ (little used)
$CdWO_4$	7.9	1325	21	8×10^{-6}	5300	γ (little used)
CaI_2 (pure)	3.96	575	460	0.55×10^{-6}	4100	γ, X
CaI_2 (Eu)	3.96	573	460	0.79×10^{-6}	4700	γ, X

solid solutions of substances possessing luminescent properties in various plastic substances. The duration of their fluorescent flash is less, by an order of 1, than that of organic crystals. They can be prepared in any desirable size, and this allows highly efficient detectors for fast neutrons to be constructed. The light yield is generally in the range of 30–65 % of that of anthracene. The light properties depend mostly on the concentration of the phosphorescent substance in the polymer [3]. At a certain concentration the light yield reaches a maximum or remains constant. Often, the scintillator contains two fluorescent substances. One of them, the so-called "wave-length shifter," allows the middle spectrum of the emitted light to be better adapted to the spectral sensitivity of the photomultiplier tube and also helps to prevent absorption in the solution.

Many of these properties are also common to organic liquid scintillators, which have acquired particular importance in the detection of very low energy beta

Table 3.2

Crystalline Organic Scintillators and their Physical Properties

Scintillator	Density	Melting point (°C)	Light yield (% in terms of anthracene)	Dieary time (sec)	Wavelength at maximum emission (Å)	Application
Anthracene	1.25	217	100	3×10^{-8}	4470	α, β, γ
Stilbene	1.16	125	50	4.5×10^{-9}	4100	α, β, γ
Tetraphenyl ($C_{24}H_{18}$)		318	90	4×10^{-9}	4400	α, β, γ,
Phenantrene	1.03	101	50	8×10^{-9}	4350	α, β, γ,
Terphenyl	1.23	213	50	4.5×10^{-9}	4000	α, β, γ
Plastic scintillators	1.03– 1.06	40 150	30 65	2.5×10^{-9}	3500 4800	all kinds
Liquid scintillators	0.7 1.6	80 270	15 80	2.8×10^{-9}	3500 4500	all kinds

radiation. These are liquid solutions of one or two fluorescent substances in suitable solvents whose nature depends on the properties of the sample being measured (see also Chapter 4.8.3, p. 203). The emitted spectrum and the light yield are related, as in the case of solid solutions (plastic scintillators), to the concentration of the dissolved substance, and likewise the light yield is similar with regard to the concentration (i.e., it either has a peak or approaches a constant value). Naturally, optimum conditions are selected in the preparation of liquid scintillators.

The most commonly employed solvents are toluene, xylene and dioxane, while the following luminescent substances are used: *p*-terphenyl (TP), 2,5-diphenyloxazole (PPO), 2-phenyl-5-(4-biphenyl)-1,3,4-oxadiazole (PBD), 2,5-bis-[5′-tert. butylbenz-oxazolyle (2′)]-thiophen (BBOT) and other compounds. The following are used as wavelength shifter: 1,4-di-[2-(5-phenyloxazolyle)]-benzene (POPOP), 2-(1-naphtyl)-5-phenyloxazole (α-NPO), 2,5-di (4-biphenyl)-oxazole (BBO), *p*, *p*′-di-phenylstilbene (DPS), etc.

The light yields of various liquid scintillators may be between 20 and 80 % of that of anthracene. The dieary time of the fluorescent flash attains a value of the order of 10^{-9} sec.

Table 3.3

Liquid Scintillators

Solvent	Scintillator (wavelength shifter)	Conc. g/l (of mano-meter)	Wavelength at maximum emission (Å)	Light yield (% in terms of anthra-cene)	Dieary time (sec)
Polystyrene	Tetraphenyl-butadiene	16	4500	36	4.6×10^{-9}
Polyvinyltoluene	TP DPS	36 0.9	3800	48	3×10^{-9}
Polyvinyltoluene	TP Tetraphenyl-butadiene	36 0.2	4450	45	4×10^{-9}
Polyvinyltoluene	TP POPOP	36 1	4300	51	4×10^{-9}
NE 102*a**			4250	65	2.2×10^{-9}
NE 104*			4050	65	1.8×10^{-9}
Toluene	BBOT	4	4350	61	5×10^{-9}
Toluene	TP POPOP	4 0.1	4320	61	5×10^{-9}
Toluene	TP α-NPO	4 0.1	4150	54	3.2×10^{-9}
Toluene	PPO	4	3820	54	3×10^{-9}
Toluene	PBD	8	3700	63	2.8×10^{-9}
Xylene	TP	5	3700	50	3×10^{-9}
Xylene	PPO	5	3800	45	3×10^{-9}
Phenylcyclohexane	TP	3	3700	50	2.9×10^{-9}
NE 213*		4250	4250	78	3.6×10^{-9}
NE 210 (gel)*			4250	50	2.3×10^{-9}

* Supplied by Nuclear Enterprises, Ltd., Great Britain.

For substances slightly soluble in the mentioned solvents, gel scintillators are used.

Some of the more important scintillators in the form of solid and liquid solutions, together with their most important properties, are listed in Table 3.3.

3.1.2.3 Photomultipliers

These are high-vacuum chambers with a built-in photocathode and a system of dynodes. Their function is to convert the light flashes of a scintillator into current pulses. There are several types of photomultiplier tubes available. A typical design is shown in Fig. 3.12. Some characteristics of the photomultipliers that may be of importance from the point of view of detection and spectrometry, are discussed below.

A basic parameter in the selection of a photomultiplier is the spectral sensitivity of the photocathode, which depends on various details in the design of the photomultiplier. Hence the photomultiplier should be carefully selected according to the scintillators used, i.e., so that its spectral sensitivity corresponds to the light spectrum emitted by the scintillator. Most routinely used photomultiplier tubes have a sensitivity of S-11, whose maximum is at 4400 Å. The spectral sensitivity of these photomultiplier tubes corresponds well with the peaks of maximum emission of many scintillators—among others, those of NaI(Tl) and anthracene, as shown in Fig. 3.13.

Another important characteristic of a photomultiplier tube is the thermal emission of electrons by the photocathode, the so-called noise of a photomultiplier, which may amount to approximately 5000 electrons cm^{-2} sec^{-1} at room temperature. Naturally, a general requirement is to reduce this noise to a minimum. An effective way is to cool the photocathodes. A drop in temperature of 10 °C causes the thermal emission to be reduced approximately by a factor of 2. It is especially important to attain a noise reduction in cases where very small pulses are produced.

Further features of a photomultiplier tube, which must occasionally be adjusted, are: the transit time of an electron from cathode to anode (must be as short as possible), amplification factor (modern photomultiplier tubes easily attain an amplification of 10^6), and operational voltage. Some of the spectrally important properties of multiplier phototubes will be dealt with in connection with gamma spectrometry (Chapter 3.1.2.5, p. 108).

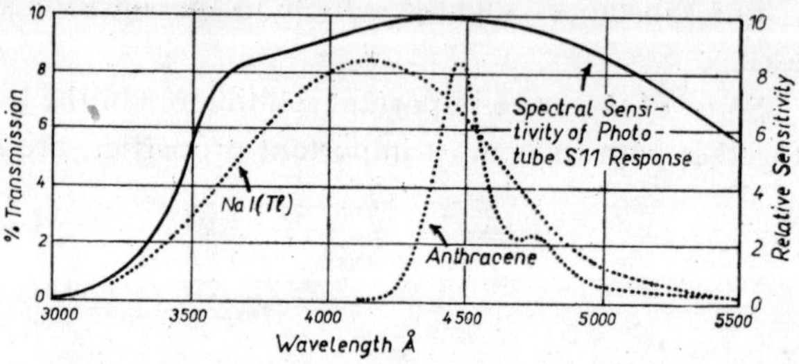

Fig. 3.13. Emission anthracene and NaI(Tl) spectra and their comparison with the spectral sensitivity of the photomultiplier tube type S-11.

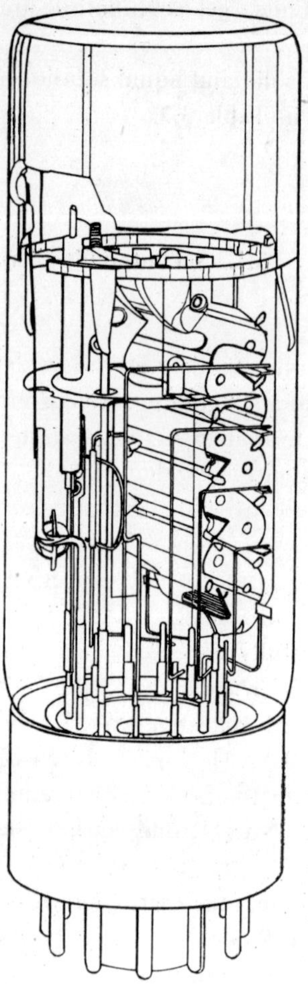

Fig. 3.12. Photomultiplier tube (Phillips).

3.1.2.4 Reflectors, Light-Guides and Arrangement of Scintillation Counter Head

It is important to have the pulse height accurately proportional to the energy dissipated in the crystal. This can be achieved only if light produced by photons or particles in all parts of the scintillator can reach the cathode with a uniform attenuation. The scintillator should be quite transparent, and nearly total reflection at surfaces other than the photocathode should be the goal. The use of HgO, TiO_2 and Al are recommended. HgO provides the best light yields; however, it is mechanically rather sensitive.

Contact between the scintillator and the photocathode of the photomultiplier tube is established either directly or by means of the light-guide. When the scintillator is placed directly on the photomultiplier tube, satisfactory and stable contact of their walls is attained with the use of vaseline and oils which are good light conductors. A silicon vaseline, for instance, is currently employed. Light conductors are gene-

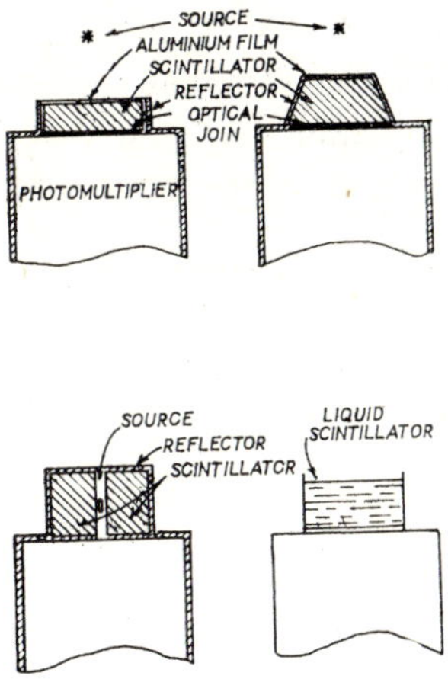

Fig. 3.14. Various ways of setting a scintillation counter.

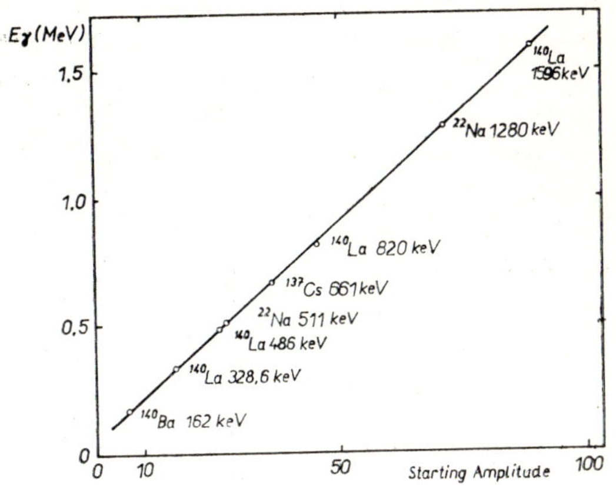

Fig. 3.15. Energy calibration of a scintillation spectrometer.

rally used only in cases where the scintillator has a larger surface than the photocathode, or when, for example, the scintillator is located in a magnetic field. In the majority of light guides, the transfer of light from the scintillator to the photocathode is based on a total reflection of light off the walls. They are prepared mostly from plastics.

Most scintillators have the shape of a cylinder, or of a blunted, cut-off cone. Also well-type forms are used (Fig. 3.14).

3.1.2.5 The Use of Scintillation Counters

Scintillation counters have a wide field of application both in routine nuclear radiation detection and in spectrometry, and the type of instrument used depends on its application.

3.1.2.5.1 Nuclear Radiation Detection by Means of Scintillation Detectors

Practically every type of ionizing radiation can be detected by means of a scintillation detector. Pulses are processed in the usual way (Chapter 3.2, p. 121) except that an integral discriminator is inserted before the recording unit, to eliminate low energy pulse noises. For universal measurements, standard scintillation counter heads are used which permit a simple replacement of scintillators and may be applied to detection of various types of radiation. Scintillation detectors are definitely superior to those based on gas ionization – particularly in gamma radiation detection, owing to their higher sensitivity, and also in the detection of very low energy beta radiation sources, by means of liquid scintillators.

3.1.2.5.2 Scintillation Gamma Spectrometry

Undoubtedly, the scintillation method finds its most significant application today in gamma spectrometry. Scintillation spectrometry is based on the proportionality between the pulse amplitude produced within the photomultiplier tube and the energy of the particle (photon) hitting the scintillator.

Increased requirements are laid on scintillators but particularly on the photomultiplier tubes for scintillation spectrometry. Amplification in the photomultiplier tube depends, in the first place, on the input voltage. For example, a 1% increase in voltage results in an amplification of approximately 7% in a photomultiplier tube with 10 dynodes. Hence, it is of extreme importance to have a highly stabilized input voltage. For purposes of spectrometry, the linear region of a photomultiplier tube should be used where amplification is independent of pulse amplitudes; thus the correlation between the energy of gamma quanta absorbed due to photoeffect, and the pulse amplitude, is also linear. When more accurate measurements should be made, the instrument should be calibrated by means of gamma sources of known energy. In addition, it should be borne in mind that pulses leaving a photomultiplier tube are subject to considerable statistical fluctuation. This is related to the yield by the photocathode (number of photoelectrons per 1 photon) and to the dynode amplification factor (number of secondary electrons per 1 primary). The photoeffect on the photocathode requires a relatively large amount of energy (of the order of 100 keV). The statistical limitation of resolution ΔE may be expressed by the relation

$$\frac{\Delta E}{E} \approx \frac{1}{\sqrt{N}}, \tag{3.4}$$

where E is the energy of the measured particles, and N the number of the elementary processes evoked by the particle.

When interpreting a gamma spectrum we must take into account its structure in certain details. All the types of interaction between gamma quanta and the scintillator (photoeffect, Compton effect and electron pair formation – see p. 64, 65) contribute to a gamma spectrum.

In the case of the photoeffect, the gamma radiation photon loses all its energy in the scintillator and the result will be a scintillation flash proportional to the photon energy. On a curve representing the spectrum there will appear a more or less clear maximum, the so-called *photo-peak* which is essentially Gaussian in character. The position and size of the peak are a source of information on the energy and intensity of the photons. Photo-peaks are a dominant component of gamma spectra.

In the Compton effect, only part of the energy is transferred to the electron. The Compton electron is stopped in the scintillator and the scattered gamma radiation can be absorbed by the photoeffect, or may leave the crystal. In the first eventuality, all of the gamma energy, which contributes to the formation of the spectrum in the form of a photo-peak, is again lost during the interaction. If no further interaction occurs, then Compton electrons alone participate in the formation of the spectrum, and their energy may range from zero to a certain maximum value, given by the relation

$$E_e = \frac{4E_\gamma^2}{4E_\gamma + 1}, \qquad (3.5)$$

where E_γ is the energy of the original gamma radiation. The result will be a wide *Compton continuum* in the gamma spectrum, energetically lower than the photo-peak. The Compton continuum is terminated before the photo-peak by the *Compton edge*, which corresponds to the energy given by the relation (3.5). If, as a result of compound gamma radiation, there is more than one photo-peak, the Compton continuum is always located at lower energies.

Similarly, the pair formation which occurs, with some probability, at $E_\gamma \gtrless 1.02$ MeV, may be involved in the spectrum formation through the energy of the created negatron, or through the energy of the photoelectron produced from photons that ensued from the annihilation process. It is thus that the *pair production peak* occurs. If a given radionuclide emits X-radiation, a photo-peak corresponds to its energy, as in the case of gamma quanta.

All these effects are illustrated in Fig. 3.16 [16]. A more detailed explanation of the shape and nature of gamma spectra may be found in various books (for example, Ref. [16–19]).

An interpretation of gamma spectra depends on the possibilities of resolving and measuring their photo-peaks and on the resolving power of the spectrometer. In order to distinguish various radionuclides in their mixture after activation, it is necessary that their photo-peaks be as narrow as possible. *The resolving power* of the given spectrometer (resolution) is usually expressed as the ratio between the photo-peak width of a standard source ^{137}Cs at half its height, to its energy (0.661 MeV), as shown in Fig. 3.17. The resolution ranges between 6 % and 12 %.

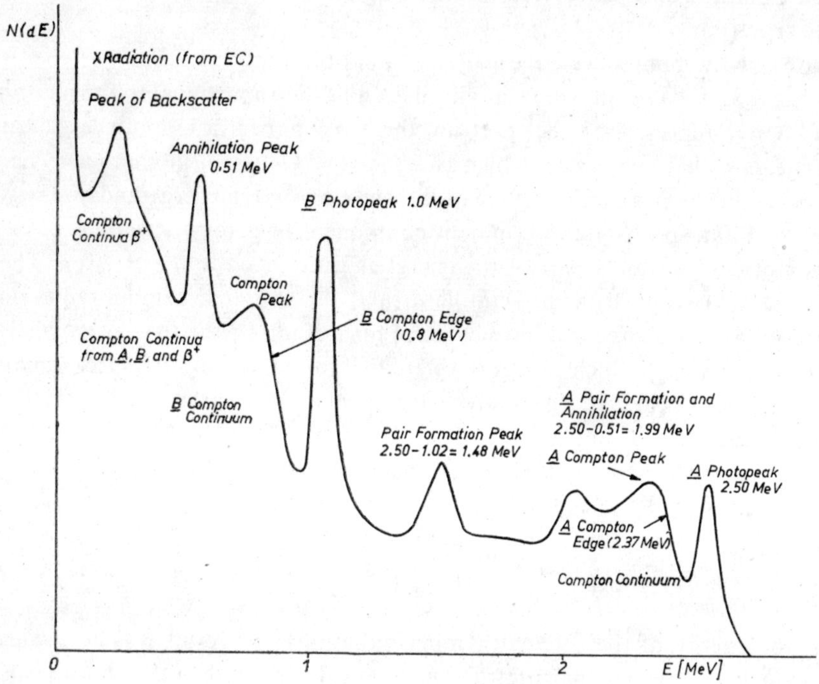

Fig. 3.16. Hypothetic gamma spectrum of radionuclide A (β^-, $E = 2.5$ MeV) and radionuclide B (β^+, EC, $E = 1.0$ MeV) [16].

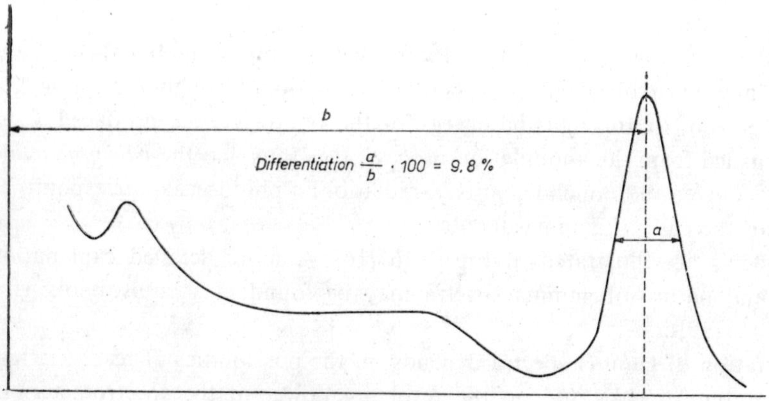

Fig. 3.17. Gamma spectrum of radionuclide ^{137}Cs.

Two peaks of equal intensity can be easily resolved if their energies differ by approximately the value of the resolving power of the measuring device [20].

Evaluation of Gamma Spectrum

The activity of a given radionuclide may be evaluated on the basis of one of its photo-peaks. Photo-peak areas or precisely delineated parts of their height can be measured. The most suitable for this purpose is the area of the photo-peak. The spectra of the investigated samples are the sum of the spectra of various radionuclides; this complicates quantitative and qualitative evaluation.

Qualitative Analysis of Gamma Spectrum. For purposes of a qualitative evaluation of an unknown gamma spectrum (for determination of an unknown energy), the spectrometer is calibrated with the aid of a radionuclide of known gamma radiation energy value. Two to three points suffice since the scale is linear. Radionuclides ^{137}Cs (E_γ = 0.661 MeV) and ^{60}Co (E_{γ_1} = 1.17 MeV; E_{γ_2} = 1.33 MeV) may be used with advantage. The data embodied in a calibration curve refer to the discriminator voltage in a single-channel or to the channel number in a multi-channel analyzer, corresponding to the photo-peak as a function of energy.

When particular radionuclides are to be identified in a mixture, another method may be used which consists of recording the gamma spectrum at certain time-intervals. According to the half-life of decay of individual photo-peaks, we may determine those that correspond to the same radionuclide, and according to their energy and half-life, we may, as a rule, identify exactly the radionuclide in question.

The so-called subtraction method is also in use: here, from a complex gamma spectrum, all its components are gradually subtracted.

Quantitative Analysis of Gamma Spectrum. For a quantitative evaluation of certain radionuclides in an investigated sample, the most frequently employed method is the relative method, in which photo-peaks of standard and the unknown radionuclides are compared. The surface of the photo-peak can usually be conveniently determined if it lies on the basic axis. Such a peak is merely the photo-peak of the highest energy. All the other photo-peaks lie on the Compton continuum, or eventually interfere with some other photo-peak. The evaluation itself of the photo-peaks may be carried out by different methods, of which the more important are discussed below.

Graphic Methods. Here we measure either the whole photo-peak, or a strictly delineated part. To assure that only the photo-peak corresponding to the given radionuclide is evaluated and not a contribution of a combined spectrum from some other radionuclide, a line is drawn extrapolating the course of the continuous spectrum under the photo-peak. The photo-peak area is measured planimetrically, or it is cut out and weighed.

Another method measures the upper part of the photo-peak above a line parallel

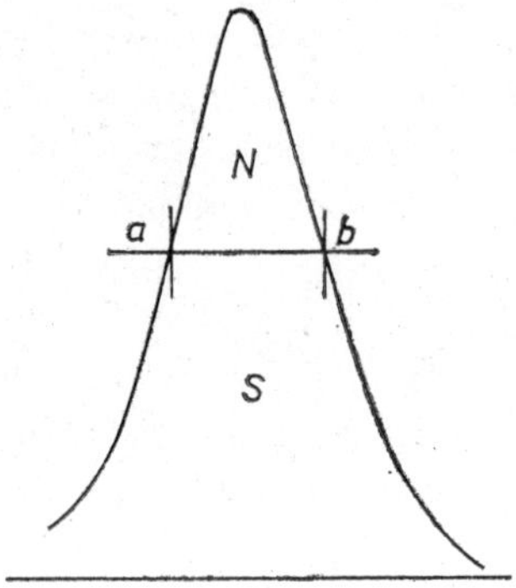

Fig. 3.18. Example of evaluation of a photo-peak of total area S.

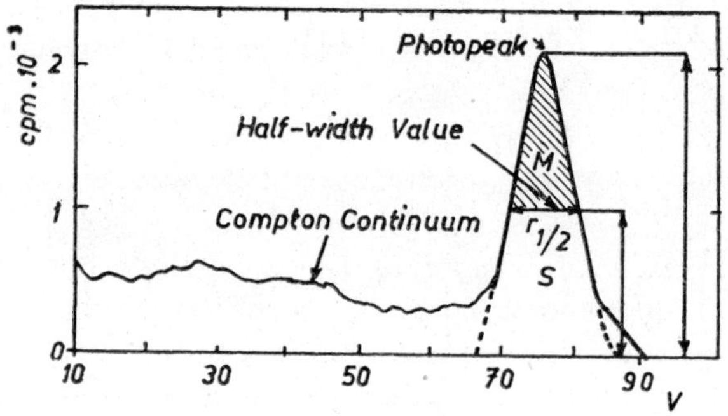

Fig. 3.19. Gamma radiation spectrum and determination of area under the photo-peak, according to McIsaac [23].

with the abscissa, of a constant length $a-b$, drawn in such a way that the ascending and the descending lines of the photo-peak curve pass through these points (Fig. 3.18). Length $a-b$ is constant for all the photo-peaks being compared. The part N of the photo-peak is proportional to its total area S and thus area N is proportional to the radionuclide activity. The drawbacks of the method are evident mainly in the measurement of high and narrow photo-peaks when the accuracy of the placing of the line is problematic, or when the photo-peak lacks a regular shape for interference with the Compton edge, or another photo-peak.

McIsaac [23] divided the area under the photo-peak into two parts by means of the value of the half-width (in parallel with the abscissa). The area above the half-width value (M) and the whole area above the photo-peak (S) are mutually related (Fig. 3.19). McIsaac determined the value M by planimetry.

Calculation methods. According to Connally and Leboeuf [22] the area under the photo-peak is given in units cpm, by the equation

$$A = \int a \, dE \tag{3.6}$$

or by approximation

$$A = \sum a \Delta E, \tag{3.7}$$

where a is the counting rate related to the unit of channel width, and ΔE the channel width in energy units.

Since the shape of the photo-peak closely resembles a Gaussian curve, this relation can be simplified. The area under the photo-peak, in cpm units, will be

$$A = sa_0r_{\frac{1}{2}}, \tag{3.8}$$

where a_0 is the photo-peak height in cpm, $r_{\frac{1}{2}}$ the half-width value in units of channel width, and $s = 1.07$, a constant derived from the Gaussian curve (Fig. 3.20).

For a correction of the Compton continuum of the component with a higher energy,

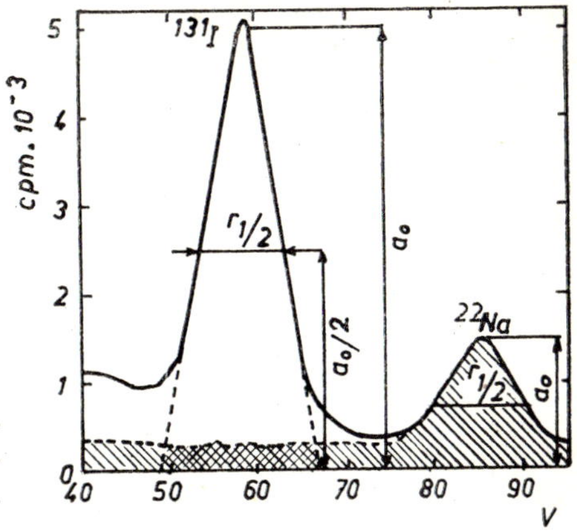

Fig. 3.20. Common spectrum of two gamma radionuclides and evaluation of area under photo-peak, according to Connally and Leboeuf [22].

Connally and Leboeuf made use of the calibration technique. The calibration curve for the given radionuclide is plotted from sample preparations of various activities from which gamma spectra are recorded. The values of the photo-peak heights obtained from these spectra then serve for the calibration curve. When several radionuclides are measured in series, the size of Compton continuum can be determined by means of the photo-peak activity and the calibration curve.

A disadvantage of the method is that accuracy in the determination of the half-width value is unsatisfactory, as the value is not derived from direct measurement. As the shape of the photo-peak differs more or less from that of the Gaussian curve, the constant in equation (3.8) is not exact.

The method by calculation as worked out by Covell [21] removes the inaccuracies arising from the asymmetry of the photo-peak. The method in its essence is shown in Fig. 3.21. The greatest number of pulses a_0 is in the channel which is in reality the "symmetry axis" of the photo-peak. The numbers of pulses in channels in the direction of lower energy are designated by $a_1, a_2, \ldots, a_n$, and those towards higher energy levels by $b_1, b_2, \ldots, b_n$. The straight line dividing the photo-peak area into two parts N

and Q, is chosen on the basis of the course of Compton continuum. The whole area delineated by the last channels on either side (a_n and b_n) are marked as P, the value of which is equal to the sum of activities measured in individual channels:

$$P = a_0 + \sum_{i=1}^{n} a_i + \sum_{i=1}^{n} b_i \tag{3.9}$$

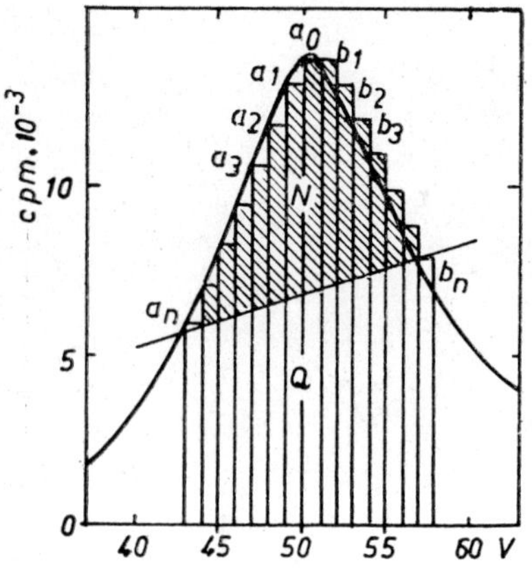

Fig. 3.21. Evaluation of area under photo-peak, according to Covell [21].

in which n gives the number of channels. The area under the straight line connecting the two extreme channels (Q) is

$$Q = \frac{(2n - 1)(a_n + b_n)}{2} + (a_n + b_n) = \left(n + \frac{1}{2}\right)(a_n + b_n). \tag{3.10}$$

Area N above the straight line, whose value is practically independent of the presence of Compton activity, according to Fig. 3.21, or equations (3.9) and (3.10), will be:

$$N = P - Q = a_0 + \sum_{i=1}^{n} a_i + \sum_{i=1}^{n} b_i - \left(n + \frac{1}{2}\right)(a_n + b_n). \tag{3.11}$$

Since it is desirable that area N should embrace a large part of the photo-peak, conditions for the determination must be chosen with great care.

The mean quadratic deviation of the value thus obtained is

$$\sigma = \sqrt{N + (n - 1/2)(n + 1/2)(a_n + b_n)}. \tag{3.12}$$

Subtraction Method. This method of quantitative determination requires the component present in the gamma spectrum, by the measurement of energy of individual photo-peaks, by measurement of the spectra in the course of decay, and by a comparison with the standard course of gamma spectra.

The first nuclide occurring in the mixture of other radionuclides is determined by means of the peak height with the highest energy. With the help of the standard spectrum of this radionuclide, this component is subtracted point by point. Again, with the help of the peak with the highest energy, the next radionuclide is de-

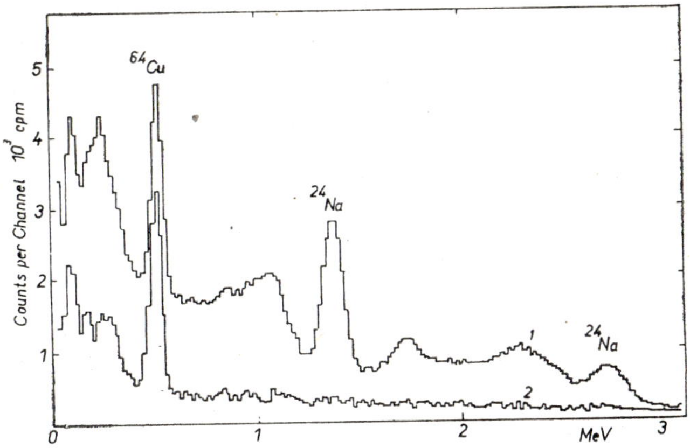

Fig. 3.22. Subtraction method. (Curve 1 — gamma spectrum of zonally cleaned aluminum, Curve 2 — the same spectrum after subtraction of ^{24}Na) [26].

termined and its spectrum subtracted. The method will thus steadily decrease in accuracy as lower energies are involved.

Many authors have worked to simplify and quicken the resolution and evaluation of spectra by this method: they used mechanical drawing devices [24], electronic subtraction [24], digital electronic computers [25] and others [20]. A subtraction carried out on a 200-channel analyzer is shown in Fig. 3.22 [26].

3.1.3 Semiconductor Counters

Recently semiconductor detectors have contributed greatly to the development of the detection technique of nuclear radiation. One of the main reasons for their rapid development is related to the problem of a statistical delimitation of energy resolution (relation 3.4). The advantages of semiconductor detectors will become evident when we consider that in applying a NaI scintillation detector, approximately 700 eV are needed to produce a single photoelectron, and in the argon-filled ionization chamber 28 eV are required for the production of an electron pair, while with silicon only 3.5 eV are necessary to produce an electron hole pair. Besides its outstanding energy resolution, it also possesses a high time-resolving capacity.

One of the characteristic properties of semiconductors is to pass, in the electric field, a certain leakage current. In order that semiconductors may be used as detectors, no current may flow through them, even when the intensity of the electric field is very high. This is achieved in two ways:

a) by the diffusion of a certain controlled amount of impurities which considerably increase resistance of the semiconductor, and

b) by the use of reverse-biased diodes.

The detectors based on the latter principle are the more significant and hence, a brief description of their mode of action will be given here.

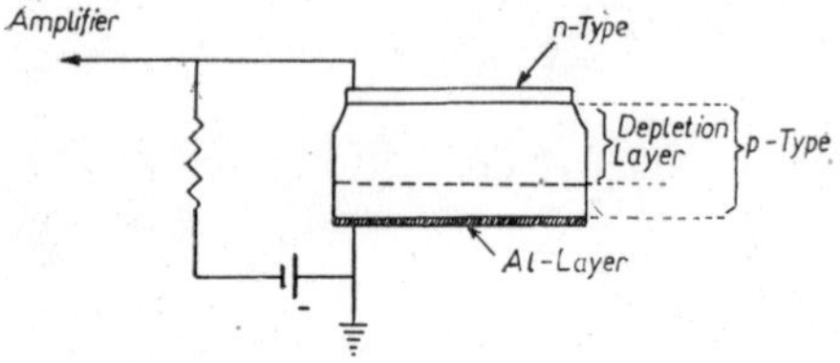

Fig. 3.23. Schematic diagram of a semiconductor *p-n* junction detector.

The basis of reverse-biased diodes is the formation of a surface barrier to the leakage current. According to the manner in which this surface barrier is formed, the detector then may be either a diffused-junction detector, or a surface-barrier detector.

1. In the diffused-junction detector a so-called *p-n* junction is produced by alloying or diffusing minute amounts of impurities into the semiconductors. The *p*-type semiconductors are characterized by the presence of an excess positive (hole) charge and the *n*-type semiconductors by an excess of the negative charge (electrons). For silicon and germanium, *n*-types may be prepared by diffusing atoms of phosphorus and arsenic (electron donors) into the surface layer of the semiconductor, under a high temperature. Boron and gallium are acceptors of electrons and yield the *p*-types.

2. In the surface-barrier detector, the barrier is produced by a suitable chemical intervention, for instance, by forming a thin coating of SiO_2 on the top semiconductor wafer, on which is an evaporated gold layer.

The principle of the mechanism of action of a semiconductor detector with reverse-biased diode will be explained in the model of a detector with a *p-n* junction, schematically illustrated in Fig. 3.23. The basic material is *p*-type silicon. A *n*-type layer is produced on one of its sides by phosphorus diffusion, while on the other, an evaporated-aluminum layer is laid to assure contact. An input voltage causes electrons in the *n*-type to be attracted towards the positive electrode while the positive holes migrate towards the negative electrode. In this way a charge-free region or, as it is also called, a depletion layer is formed. It is precisely this layer that forms the sensitive volume of a semiconductor detector. When an ionizing particle enters this layer, it creates in its path negative electrons and positive holes which rapidly migrate towards the opposite electrodes. This charge displacement is already analogous with that of ions in ionization chambers. A current is thereby produced on the outer circuit, which becomes converted on the resistors into a charge pulse, and the latter may further be shaped and recorded. The charge thus created, and hence also the pulse, is directly proportional to the particle energy, and this makes it possible to apply semiconductor detectors in the spectrometry of nuclear radiation. This is the principal manner in which semiconductor detectors are used, for in routine measurements of activity we prefer, for the time being, other detection methods.

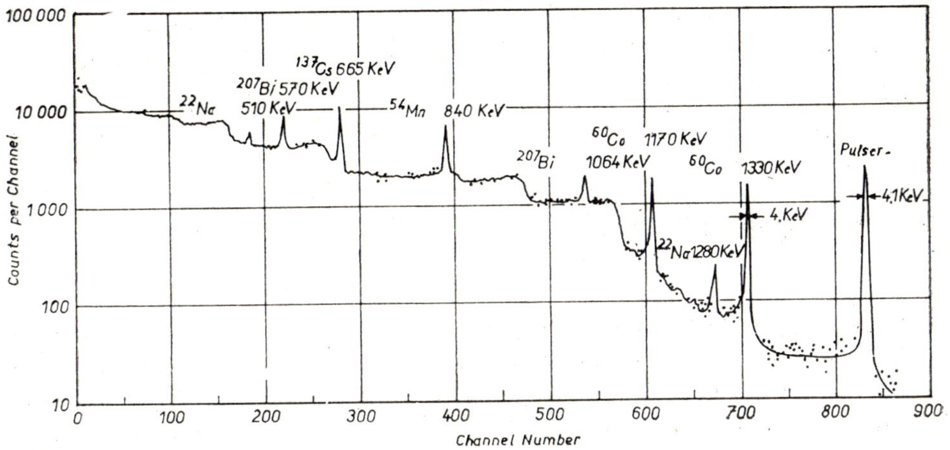

Fig. 3.24. Spectrum of radionuclide mixtures recorded by means of a germanium detector.

A drawback of semiconductor detectors is that they can be prepared only with a relatively small thickness. However, the condition is valid here also that when an energy spectrum is being recorded, radiation is completely absorbed in the detector. This requirement could be fulfilled so far only in the case of charged particles where the energy-resolving power amounts to 0.2 %. Recently, however, semiconductor detectors have been successfully employed also in gamma spectrometry. The thickness of the sensitive layer of a detector has been effectively increased to such an extent that electrons produced by the photoeffect of gamma radiation are completely absorbed. Thus, for instance, ^{137}Cs and ^{60}Co gamma spectra have been obtained with a resolving power better than 0.5 %. An example of a spectrum obtained by a semiconductor detector is shown in Fig. 3.24.

The amplitude of the initial pulse from a semiconductor detector is but slightly dependent upon the input voltage. Therefore, the required pulse size cannot be set by a change in voltage as is the case in proportional and scintillation counters.

At the present time silicon and germanium detectors are in use, both with diffused junction and with surface barrier; more recently lithium has been added to the silicon and germanium to give the so-called lithium drifted detectors. Germanium detectors have the disadvantage that they operate only at low temperatures. Without lithium they operate at the temperature of liquid air; lithium drifted detectors operate at about −70 °C. From this point of view silicon detectors are more suitable as they can be used even at room temperature—hence they are preferred for spectrometry of charged particles. For gamma spectrometry, germanium detectors are better suited, despite the necessity of cooling; because of their higher atomic number, they are more efficient for the photoeffect. In addition, the energy requirement for the production of an electron hole pair is smaller (2.8 eV) than with silicon (3.5 eV).

For more detailed information on semiconductor detectors we refer the reader to the literature [12, 13].

3.1.4 Photographic Method

The interaction of all the types of radioactive radiation with a suitable photographic emulsion results in a latent picture. The mechanism of its origin is similar to that with visible light. This effect may be *collective* (when the intensity of the radiation source is judged according to the overall blackening), or *individual* (when the paths of individual particles are pictured). The collective effect is of greater practical significance in the application of nuclear radiation to chemical analysis. This collective effect is judged according to the blackening, which can be quantitatively evaluated, for instance, with a photometer. Blackening is defined as the logarithm of the ratio between the intensity of the incident light upon the plate and the intensity of light that had passed through the emulsion and the plate. Occasionally, the collective effect is evaluated by counting the number of developed silver grains under a microscope.

A basic requirement in the determination of activity by this method is the homogeneity of the sample. The latter is suitably put into contact with the emulsion. In the case of radioactive solutions a sample drop may be transferred on the emulsion where it is allowed to dry, or the emulsion is immersed in the solution where the gelatin swells and then is desiccated. Another technique consists of wrapping a thin capillary, which contains the radioactive solution, with the liquid emulsion. Solid samples are placed with their smooth surface on the photographic plate; often a thin layer of "zapon" lacquer is introduced between them. The exposure time depends on the type of radiation, sample activity and emulsion sensitivity.

Photographic methods in radiochemistry or nuclear analytical chemistry are used for the following purposes:

a) to determine sample activity and concentration of radioactive solutions according to the degree of blackening of the photographic emulsion under the action of radiation for a certain period of time,

b) to measure radiation doses (film dosimeters),

c) to record the distribution pattern of radioactive nuclides (on paper radio-chromatograms, in alloys, in precipitates, etc.).

The individual effect (microradiographic method) may be utilized for

a) identification of radionuclides by measuring the particle range in the nuclear emulsion,

b) determination of radionuclide concentration, based on the known decay constant,

c) determination of the decay constant according to the absolute number of traces in the emulsion, produced by a known amount of the radioactive substance per unit of time,

d) evaluation of the radioactive purity of the sample,

e) tracing nuclear reactions, etc.

The photographic method is very sensitive, and relatively simple. Its results are visible and convincing. It does not require any complicated electronic instrumentation. However, great accuracy, particularly in the case of the collective radiation effect, cannot be achieved by it. It yields reliable results only in qualitative studies. More detailed information on the photographic method is supplied in specialized treatises [27, 28, 29].

3.1.5 Measurement of Neutron Radiation

Neutrons, in contrast to charged particles, and gamma and X-radiation, are incapable of interaction with the electron shell of the atoms, and consequently do not produce in it any ionization or excitation. Their detection is based on the ionizing ability of products that result from the direct interaction between neutrons and the nuclides contained in the detectors. Here, the following events may be involved:

a) elastic scattering on light nuclides,

b) capture transmutation reactions (n, α) and (n, p),

c) capture activation reactions (n, γ),

d) nuclear fission reactions (n, f).

The scattering intervenes particularly in the detection of fast neutrons. Advantage is taken here of the fact that neutrons, when colliding with nuclei, transmit to them part of their kinetic energy (in central collisions they transmit most 50% of their kinetic energy). This gives rise to recoil nuclei which may be registered as charged particles when their energy is sufficiently great. The number of recoil nuclei is proportional to the number of neutrons which passed through the effective volume of the measuring device, and their energy, too, is proportional to that of the neutrons. This means that by measuring the number and energy of recoil nuclei we can determine the intensity of flow and the energy of neutrons. Since the energy of the recoil nucleus also depends

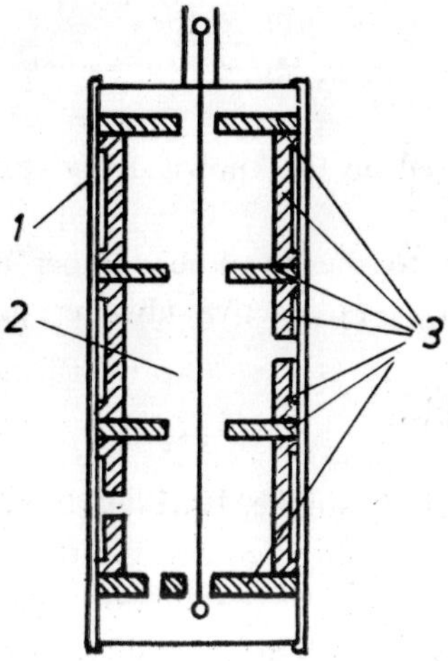

Fig. 3.25. Schema of a proportional counter of fast neutrons (1 — cylindrical cathode, 2 — fiber anode, 3 — polyethylene).

on the mass ratio of neutron to nucleus, a proton (hydrogen nucleus) can acquire the greatest energy, because it has the smallest mass of all nuclei. Detection of fast neutrons on this principle is carried out mostly by means of ionization chambers, proportional counters and Geiger-Müller counters which are filled with hydrogen or with hydrocarbons, namely methane, ethane, propane, butane, ethylene or ammonia under suitable pressure.

Methane- and argon-filled proportional counters are often used to detect fast neutrons. Polyethylene "proton radiators" are installed between the cylindrical cathode which is simultaneously the wall of the tube, and the fiber anode. In their passage through the polyethylene barrier, fast neutrons knock out protons from the hydrogenous material, which then ionize the tube filling. A scheme of this type of counter is shown in Fig. 3.25 [30]. The detector is insensitive to alpha and beta radiations because these do not penetrate the wall of the tube, and with suitable discrimination, pulses due to gamma radiation are not recorded. Also, slow neutrons cannot be detected; hence, such a counter is specific for fast neutrons.

Fast-neutron detection based on this principle may utilize nearly all the types of plastic and liquid scintillators which contain hydrogen in high percentage. The use of scintillators as detectors (and also proportional counters) offers the advantage that by discrimination, we may eliminate pulses caused by other types of radiation.

For the detection of slow neutrons, a nuclear reaction is most frequently used that leads to the emission of charged particles, for the latter can be recorded simply and with high efficiency. Boron (^{10}B) and lithium (^{6}Li) are the most suitable. The inner walls of the detector cathode (in the ionization chamber, proportional or GM counter) are coated with pulverized boron, its compounds, or compounds of lithium. A more convenient way is to fill the detector with gaseous BF_3. Under slow neutron irradiation, the following nuclear reactions take place: ${}^{10}_{5}B(n, \alpha)\, {}^{7}_{3}Li$ and ${}^{6}_{3}Li(n, \alpha)\, {}^{3}_{1}H$. The emitted alpha particles ionize the gas filling and thus enable neutrons to be recorded. The cross-section of these reactions increases with decreasing neutron velocity. That is why these nuclear reactions are employed for slow neutron detection.

The same reactions may be advantageously utilized for the detection of slow neutrons by means of a scintillation counter. For example, scintillator ZnS(Ag) contains a compound of lithium enriched in ^{6}Li, or a compound of boron which contains, without enrichment, ^{10}B in 18.5 % abundance. Such scintillators enable neutron radiation to be detected even when 10^7 gamma quanta corresponds to one neutron. Boron-containing plastic or liquid detectors may similarly be used. The scintillation method is one of the most advantageous for the detection of slow neutrons.

Slow neutrons may further be detected through nonelastic scattering when the excited nuclei emit gamma quanta due to by capture reaction (n, γ). Gamma radiation is detected here.

As in the case of boron and lithium, the fission of heavy nuclei, e.g., uranium 235 or plutonium 239, may also serve for the detection of neutrons. The cathode of the detector is coated with one of these substances. The interaction with slow neutrons produces nuclear fission and the fission fragments carry several charges and particles which cause a strong ionization. The value shown by the detector, the number of pulses, is proportional to the flux of neutrons.

The neutron flux may further be measured by the radioactive indicator method. This involves neutron capture in which a radioactive isotope is created. The amount of the accumulated radioactive isotope is proportional to the neutron flux and can be determined by measuring its activity.

3.2 Pulse Recording and Analysis

Pulses caused by individual particles or photons of nuclear radiation in Geiger-Müller counters, pulse ionization chambers, proportional counters, scintillation counters, etc., may be recorded after amplification, either individually, by means of electronic devices which are generally called scalers, or as a whole, by measuring their integral value, using electronic equipment called counting-rate meters.

If a counter tube operates in the proportional region, or if a scintillation counter is used, there ensues the task of analyzing the pulses according to their height—this is done by means of the so-called pulse-height analyzers. Working with certain types of counters often requires recording of pulses in coincidence or in anti-coincidence.

3.2.1 Pulse Amplification

Signals obtained in radiation pulse detectors are generally very weak. In order to transmit the voltage pulse leaving the detector to the counting and integration circuits, or the coincidence circuit and amplitude analyzers an amplifier is mounted directly behind the detector. In specific cases its role is not to amplify signals, but to adjust the output impedance of the detectors to the input impedance of the measuring instruments.

In electron tube amplifiers, pulses from the detector are transferred to the electron grid, with a working resistance coupled to its anode circuit. Amplification of a device is given by the ratio of the output to the input pulse amplitude. Generally, several electron tubes in the amplifier are so connected that the total amplification is the multiple of amplifications of individual tubes.

Scattered capacities and electrode capacities of the electron tube cause a distortion of the pulse. When measurements are made with scintillation detectors, linear pulse amplifiers are used, which cause only a slight deterioration of the pulse shape. Pulses from Geiger-Müller counters may be amplified by means of a simple one-step amplifier, for they have a rather large amplitude (0.3–10 V).

To ensure stability of amplification of an amplifier, part of the output voltage is fedback to the amplifier input in such a way that it acts against the effect of the input voltage (the so-called negative feedback). In a multi-step amplifier, amplification must be so adjusted during measurement, so as to prevent excitation in which the output pulse amplitude is lower than that of the input pulses and amplification, with a simultaneous deterioration in the resolution ability. This is particularly important in spectrometric measurements. Consequently, amplifiers destined for spectrometric measurements with proportional counters have built-in circuits that limit the possibility of excitation. Linear pulse amplifiers are also constructed with transistors [31].

Electronic devices have, besides amplification, additional functions, such as pulse phase inversion, resolution of pulses of various heights (discrimination) and pulse shaping.

3.2.2 Voltage Sources for Radiation Detectors

Various high-voltage sources are used to feed radiation detectors. The choice of a suitable source is guided by the type of detector. Stability of voltage is of greater importance in nuclear instruments than in any others, for small changes in voltage may cause considerable errors even in the best counters. A measure of the quality of a voltage source of a detector is its stability, which is generally expressed as the ratio of the size of change in the output voltage following stabilization to that in the input voltage prior to stabilization.

Geiger-Müller counters operate on sources in which the output voltage remains stable within a range of $\pm$ 1 %. Scintillation counters require a stability at least 10 times higher, particularly if there is question of feeding a spectrometric sonde.

3.2.3 Scalers

Various types of recording devices may be used for registering pulses. In the simplest cases an electromagnetic and fast mechanical recorder may be applied. Such a recorder, however, possesses a low resolving power and at a higher counting rates (approximately above 400 cps) is too slow to record pulses.

The resolving power has been most successfully enhanced in counting devices, called scalers. They combine electronic computing with mechanical counting. Scalers may be binary or decadic.

Binary scalers have electronic circuits that produce one output pulse for every two input pulses. Hence one such stage halves the number of pulses. A sequence of n such stages produces a number of output pulses 2^n-times lower than that of the input pulses. The significance of scalers is that the mechanical recorder at the output end records not every pulse but, in a one scaler stage, every other one; in a sequence of two scaler stages, every fourth; in a sequence of three scaler stages every eighth; in a sequence of n scaler stages, every 2^nth. A sufficient number of scaler stages achieves the required resolving power, without the necessity of using an extra-fine recorder. All of the scaler stages carries signal neon tubes by means of which the number of pulses may be determined with an accuracy of ± 1 pulse. Thus, for instance, in a sequence of six scaler stages, the scaler transmits through, output into the mechanical recorder, only every 2^6th, i.e., every 64th pulse. This implies integer multiples only. Individual pulses whose number is lower than 64 may be read directly from the signal neon tubes.

Decadic scalers are employed more widely. Here, the results are read off directly in the decimal system. The first scaler stage records 1–9 pulses, the second 10–99, the third 100–999. These first three stages are usually formed by decades containing neon tubes or decade counter electron tubes ("*Dekatron*"). Values of the order of 10^3 pulses are carried over to the mechanical recorder where they are recorded. If, for example, the number of recorded pulses is 11,523, the mechanical recorder will show 11, the third decade 5, the second 2 and the first 3.

Lately, special counting electron and discharge tubes have been developed which considerably simplify the construction of decade counters or scalers.

Dekatron is the trade name of one type of "cold-cathode counter tube", or "decade glow counting tube", also called simply counter tube or counting tube. This is a multi-electrode, gas-filled electron tube whose function is based on the glow discharge principle. If the Dekatron is filled with neon, then it is a multi-electrode glow-lamp.

Only one cathode "shines" in the Dekatron (in reality only the gas in the vicinity of the cathode is aglow), and the second cathode may be "lighted" only when the first has gone out. Ions produced by residual ionization concentrate at a distance of some tens of microns from the cathode surface within about 10 μsec following interruption of its glow. If a signal reaches the neighbouring cathode within this interval, these ions pass towards it and the cathode is set aglow.

If the Dekatrons are set in series, an arbitrary dividing ratio may be obtained. The counting system with Dekatrons is more convenient and more compact than that with neon glow-lamps. The principal limitation of Dekatrons is in their low counting speed which as a rule, does not exceed, 10,000–20,000 cps.

Of great significance and widely used are electron counter tubes produced by various manufacturers under the trademark E1T (ribbon-beam tube). This consists essentially of a small fluorescent screen on which a flat beam of electrons deviates along a circular "step" from one side to the other. If pulses are transmitted to the left deflection-plate, the electron beam gradually takes up one of the ten permanent positions which correspond to the numerals 0 to 9 in the decimal code. Every recorded pulse shifts the electron beam by one place from zero towards nine. At the tenth pulse the electron beam returns to the initial zero position and the counter tube automatically sends a pulse into a further, similar electron tube.

In addition to counting, this electron tube performs also the task of an indicator. On its bulb there are two rows of numerals from 0 to 9. The electron beam falls on the bulb wall, coated inside with a luminescent substance. This substance glows constantly at the site of the incidence of the flat electron beam and thus marks about the relevant numeral.

A great advantage of the electron counter tube E1T resides in the fact that it requires only simple electronic circuits for its operation with a single electron tube and a minimum number of accessories. This allows very simple construction of counting decades and of complete scalers which have proved to be very reliable.

The electron counter E1T counts with a speed of 30,000 to 50,000 cps.

In the case of a *Trochotron* (or beam-switching tube) the electron beam is deflected in a circle by means of a permanent magnet. The shift from one stable position to another is made electrostatistically. It may be used up to a speed of 300,000 to 400,000 cps.

The resolving time of a scaler (the shortest period between two pulses which are counted) is usually about 50 μsec. If the pulses follow one another at shorter intervals, they are counted as one pulse. For measurements with scintillation detectors, counters with a higher resolving power are used (10^{-5} to 10^{-7} sec).

The accumulation ability of a scaler is given by the number of scaler decades and the mechanical recorder. For current measurements it is enough if the scaler is capable of recording a total of 10^5–10^6 pulses.

3.2.4 Counting-Rate Meters

In the measurement of samples with higher counting rates and when the claims on accuracy are lower, scalers may satisfactorily be replaced by the so-called counting-rate meters in which the average counting-rate, determined by the time constant, is transmitted to a milliamperemeter pointer. The milliamperemeter scale is suitably calibrated, e.g., in cpm.

Rate meters are based on the following principle: pulse size in the Geiger region depends on electrode voltage, the nature of the filling material, and dimensions of the counter. When these are kept unchanged, individual current pulses may be considered to be equal. Under such circumstances, the mean value of the current flowing through the counter is proportional to the number of current pulses and therefore to the number of particles that have reached the discharge space of the counter. To measure the mean value of the current, a RC circuit (resistance capacitor circuit) with a high time constant is connected in parallel to the measuring device, itself integrated with the anode circuit of the end-electron tube on whose grid the pulses are conveyed. As a rule, the RC value can be altered so as to permit the time constant to be adjusted to the counting rate. When lower counting rates are involved, we generally use a higher time constant so as to rule out statistical deviations of the pointer under the effects of individual pulses.

However, the height and shape of the pulses produced in Geiger-Müller counters depend, in reality, also on the time elapsed since the last pulse. If the counter is still in the recovery stage, i.e., all the positive ions have not yet reached the cathode, then the pulse provoked by a new particle is, under these conditions, smaller than that which follows complete recovery. Therefore, a special assembly must be used: an electric cell, making the pulses uniform in size and shape, is inserted before the integration end-electron tube.

In another widely-used circuit, the RC cell is charged from a double diode, or duodiode with normalized pulses (the so-called diode pump – an integrator with a dosage condenser). This arrangement is applied in particular when precision measurements in counting rates are required. Another simple but very accurate measuring device is the integration circuit with a thyratron.

With the help of rate meters we can rapidly obtain approximate information about radiation intensity. Advantage is taken of this possibility, before all else, when rate meters are used in working operations for continuous noncontact measurements made on the basis of absorption and scattering of radioactive radiation, and in various automated continuous analyzers.

3.2.5 Choice of the Most Suitable Measuring Device

The selection of the most suitable measuring device is conditioned by degree of precision required, reliability, and degree of automation. A block scheme of the overall set-up is given in Fig. 3.26.

The reliability theory on electronic instrumentation has confirmed that when instruments fulfilling the same function are mutually compared, the highest failure rate is shown by the one which has the largest number of electron tubes and semiconductors. Consequently, this should be borne in mind when an electronic measuring device is to be set up.

No universal rules can be laid down for the selection of the most suitable measuring equipment. Universal measuring sets have been found convenient for working places with minor-scale operations. Here the time losses due to the adjustment of the instrument for different measuring methods are offset by the wider applicability range of the instrument. In the larger type of operation, on the other hand, single-purpose simple instruments with a maximum degree of automation are better suited for repetitive measurements. Failure rate in these instruments is very low [31].

3.2.6 Coincidence and Anticoincidence Circuits

A whole series of measurements in nuclear analytical chemistry employs *coincidence circuits* which record only those pulses that are produced in two or more counters simultaneously. Hence, a coincidence connection permits determination of the simultaneity of two or more events. Both the GM and the scintillation counters may be used for this purpose.

Two detectors, each of which is sensitive only to a certain type of radiation, are placed close to the radioactive source as shown in Fig. 3.27. Detector *I* records beta

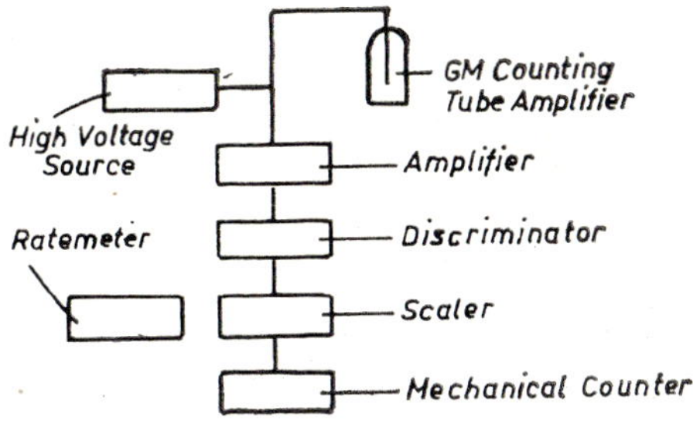

Fig. 3.26. Block diagram of connections in a rate-meter.

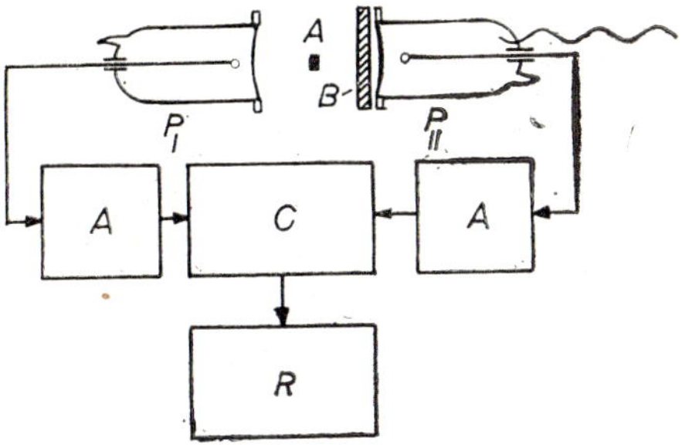

Fig. 3.27. Schema of coincidence connection (A — source, B — absorber, P_I, P_{II} — GM counters, A — amplifier, C — coincidence equipment, R — recording device).

particles; beta particles do not penetrate into detector *II*, for they are absorbed by the absorber, while gamma quanta are recorded in it through the intermediary of secondary electrons which are created by their action in the absorber. Whenever beta particles and gamma quanta—emitted simultaneously—hit their respective detector, the electronic device with which both the counters are connected, records the signal. In the case of all the other pulses which are not simultaneous, the coincidence circuit makes no recording.

The coincidence method is likewise employed for the determination of the direction of flight of a high energy particle, or for the selection of particles which follow a given direction. On this principle are built counting telescopes used for investigating cosmic rays, and also in some tests with accelerators. Figure 3.28 shows all three counters connected in coincidence, hence, they record only that particle which has simultaneously passed through all three. This allows the direction of flight of the particle to be determined.

The simultaneity of two events cannot be ascertained to perfection, for the detectors do not react instantaneously. Their response may differ by a certain maximum time interval, which is known as the resolution time. The coincidence method is the more accurate; the resolution time is shorter.

We differentiate between true and false coincidences: false coincidences are caused by various pulses produced, e.g., by background, in the detectors virtually simultaneously. To reduce the number of false coincidences, we reduce the resolution time of the circuit. With a reduction in the resolution time, the number of false coincidences will decrease, until at a certain value, they will decline practically to zero.

Anticoincidence circuit is a special type of coincidence circuit and differs in that it registers pulses from one group of detectors on condition that they are not simul-

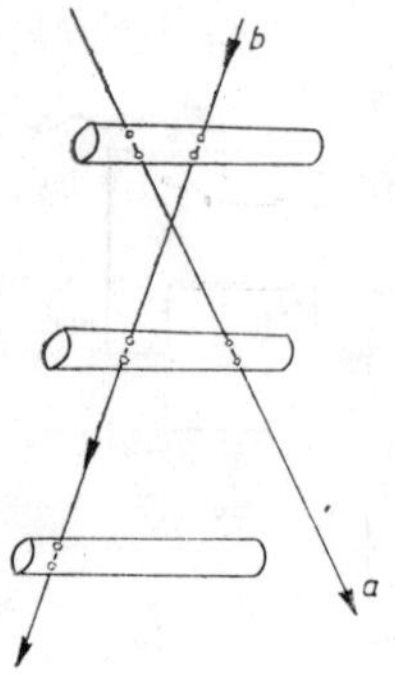

Fig. 3.28. Three counters in coincidence connection for ascertaining direction of particle flight. (Particle *b* alone may start all three counters simultaneously and register a pulse; particle *a* which starts two counters, is not registered.)

taneously registered by the second group of detectors. Anticoincidence circuits find an application, for example, in pulse height analyzers, or for reducing the detector rest pulses (the so-called anticoincidence screening).

While the coincidence circuit allows a substantial reduction of the background counts caused by noises, the anticoincidence circuit enables that portion of the background to be reduced which is due to cosmic radiation. By combining the two circuits, we may reduce the background counting rate down to 1 cpm and lower.

The coincidence method may generally be used with advantage in nuclear analytical chemistry for gamma spectrometric measurements *(coincidence gamma spectrometry)*. This is based on the phenomenon that, when an atom disintegrates, two gamma quanta, or a beta particle and a gamma quantum, are emitted simultaneously or immediately following each other, or, in the case of positron radiation sources, a coincidence of 0.51 MeV annihilation photons takes place. In view of this, three principal types of coincidence gamma spectrometry may be considered:

Gamma-gamma coincidence occurs when, following the emission of a gamma quantum of energy E_{γ_1}, the nucleus passes to the transient energy state and immediately emits a further gamma quantum of energy E_{γ_2} (e.g., in the case of radionuclide ^{46}Sc). The technique of gamma-gamma coincidence spectrometry and its application in activation analysis has been dealt with by Ljunggren [32, 33]. His experimental arrangement involved two scintillation detectors integrated as shown in Fig. 3.29. The spectrometer records pulses only when the pulse in the first detector corresponds to the given gamma radiation energy (E_{γ_1}); that in the second corresponds to the given energy E_{γ_2} and if both pulses occur practically simultaneously.

Beta-gamma coincidence takes place when the nucleus emits in immediate succession a beta particle and a gamma quantum. This type of spectrometer uses a beta and a gamma detector integrated in coincidence coupling. The spectrometer registers pulses only when the pulse evoked in the gamma radiation detector is practically simultaneously followed by a beta radiation pulse in the second detector. This techni-bue is employed when the maximum beta particle energy of the measured radio-

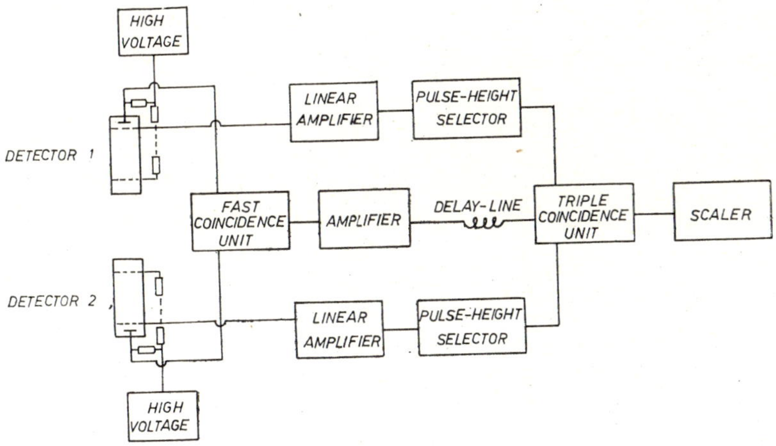

Fig. 3.29. Block diagram of a coincidence spectrometer [32].

nuclide far exceeds that of the interfering radionuclide, whereby the two can be differentiated. This technique has been used to enhance the detection selectivity of radioactive manganese in the presence of ^{24}Na [34].

Coincidence of 0.51 MeV annihilation photons is utilized in the detection of a given positron-emitting radionuclide when there are no other positron-emitting radionuclides in the sample. This type of coincidence gamma spectrometry is used in copper determination by measurement of the radionuclide ^{64}Cu [18].

3.2.7 Pulse-Height Analyzers

In nuclear chemical analyses it is often necessary to ascertain the energy and intensity of the radiation source. This requires the use of a detector which produces pulses proportional to the energy of the radiation source and these pulses, after amplification, are to be sorted according to their amplitudes. One way is to record these pulses in a loop oscillograph on photosensitive paper and then to sort them according to the size recorded on the paper. At the present time, however, pulses are sorted exclusively with the aid of pulse-height analyzers in which pulse distribution is done through an integral or a differential discriminator. The pulse-height discriminators serve to differentiate individual electric pulses from the detector according to their height.

In an *integral pulse-height analyzer* (the so-called pulse-height discriminator)

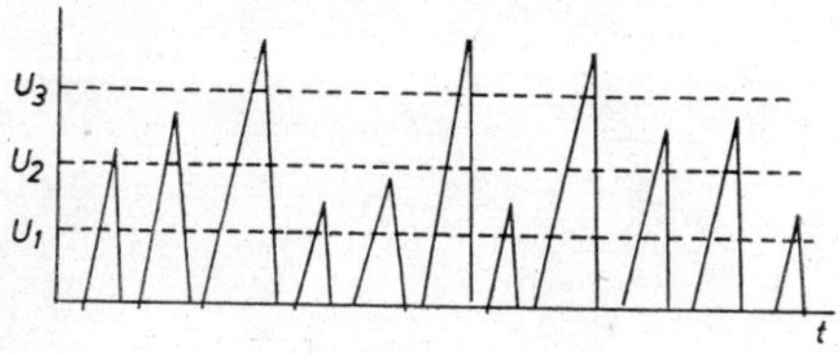

Fig. 3.30. Principle of discrimination by means of an integral discriminator.

a signal appears on the output only when the pulse-height of the input signal exceeds a pre-set discrimination level. The pulse-height spectrum shown in Fig. 3.30, is obtained by means of the integral discriminator. In the discriminator's voltage position U_1, every signal reaches the counter; in the position U_2, only seven, and in position U_3, only three signals do so.

An integral pulse-height discriminator has a linear characteristic, which means that a pulse appears at the output circuit only when the signal pulse-height exceeds a certain minimum value (the so-called amplitude level, or the threshold value). The amplitude level is, for the most part, variable, so that by raising it, we may gradually reduce pulses that have a lower pulse-height. For their discrimination we employ circuits in which are connected electron tubes or semiconductor elements. The discrimination level may be altered by means of a potentiometer, or a stepping-relay.

Integral pulse-height analyzers are used in cases where the amplitude distribution is not followed specially, that is, when only radiation is to be detected. In such a case, with the help of the pulse-height detector we "shut-off" the undesirable noise (e.g., the noise of a photomultiplier in gamma radiation detection, or background due to gamma radiation in the detection of neutron radiation, etc.).

In a *differential pulse-height analyzer* the signal at the output appears only from those input pulses whose pulse-height lies between two threshold values, U_{min} and U_{max}. The difference $U = U_{max} - U_{min}$ is called *channel width* of the pulse-height analyzer. Pulses, first amplified by a linear amplifier, are conveyed simultaneously to the input of two discriminators with a different threshold voltage (Fig. 3.31). If the pulse is higher than the threshold of both the discriminators, then it passes through both channels; the anticoincidence circuit fails to register it. If the pulse is lower than the level in both the discriminators, then it fails to pass through the channels and is not registered. And lastly, if the pulse size lies between the voltage sizes of both the threshold values of discriminators, then the pulse will pass through only one of the channels. An anticoincidence event takes place and the anticoincidence circuit records the pulse. Figure 3.32 illustrates schematically the working principle of a differential pulse-height analyzer.

Pulse-height analyzers are an indispensable part of nuclear radiation spectrometers. Simpler arrangements employ a single-channel pulse-height analyzer (a single-

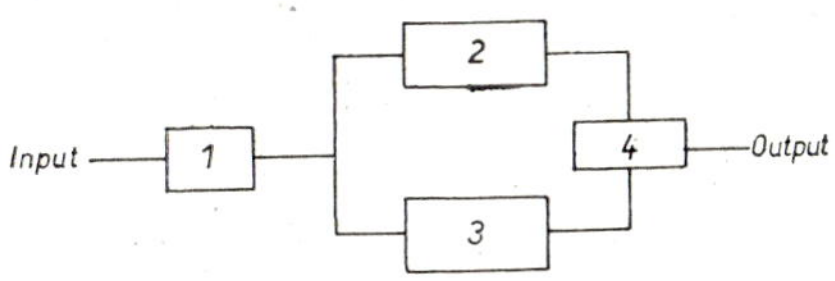

Fig. 3.31. Block diagram of a differential pulse-height analyzer (1 — amplifier, 2 — discriminator with a high threshold 3 — discriminator with a low threshold, 4 — anticoincidence).

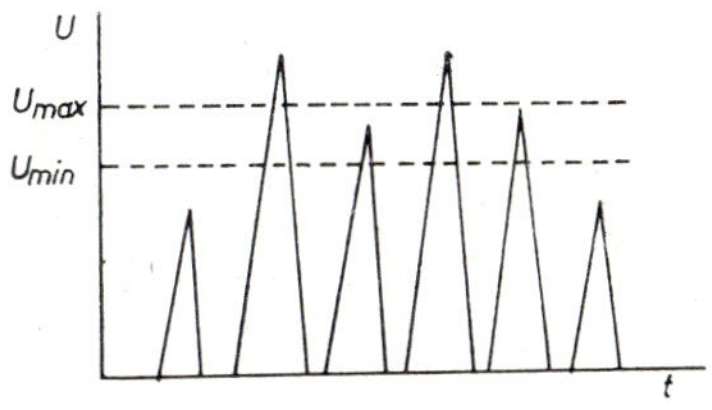

Fig. 3.32. Principle of discrimination by means of a differential amplitude analyzer.

channel spectrometer), while in other cases, multi-channel analyzers are used (multi-channel spectrometer).

In *single-channel analyzers* the spectrum is obtained when a channel of desired width is shifted in such a manner that the instrument gradually records various values of the voltage pulses, from the lowest to the highest.

Single-channel analyzers are suitable for measuring radionuclides with a half-life of 10 hours or more. The activity of the measured radiation sources in this case should be higher than 10^5 decays per minute, because the entire energy spectrum is gradually measured [16]. An improvement of these arrangements has been brought about with an automatic measurement of the whole spectrum and recording on the instrument.

The measurement with single-channel pulse-height analyzers is rather time-consuming, for the counting rate per minute has to be determined for every discrimination level. Hence, efforts have been made to measure the counting rate in all the discrimination levels simultaneously. This assures a considerable shortening of the measuring procedure, for the time taken to measure the counting rate at one discrimination level suffices for measuring that of all the levels. The arrangement making this possible is called a *multi-channel pulse-height analyzer*.

Numerous types of these analyzers have been constructed. The simplest of them were prepared by dividing the selected working analyzer band into a certain number of sectors, for example 10, so that each sector formed essentially a single-channel pulse-height analyzer with a firm discrimination level of a definite width. The discrimination levels are set to cover a certain voltage range. To be registered at the set discrimination range of the channel, a pulse must reach one discrimination level, but not the next higher one. Only such a pulse can be registered in the designated channel and is recorded by the counter in the same manner as in a single-channel analyzer. From this it is evident that, for instance, for a 10 channel pulse-height analyzer, we must have 10 single-channel analyzers and read off the values equally from 10 scalers. Therefore, this analyzer is unsuitable when a large number of channels are involved.

More advantageous is an analyzer provided with a memory unit in which the pulse-height for every pulse is determined and a corresponding channel for this pulse-height is opened by the memory unit, so that pulses with equal height are summarized in the same channel. The *memory* unit may by either *dynamic* (i.e., the record in the memory unit has to be continually repeated and is effaced when the instrument is switched off) or *static* (when the recorded value is fixed in the memory unit, even after switching off of the instrument, and may be evaluated when the latter is switched on again). The data from the memory are conveyed to a screen so that as soon as they are encoded in the memory unit, we may see the measured spectra. Permanent recording of spectra is done either by a printing system which prints the number of the channel and the number of pulses in the channel, or by a recording device which draws this relation graphically. The permanent recording is done in a very short delay, in 2–3 min. The spectrum may likewise be recorded on a tape and evaluated in an electronic computer.

The design and the number of channels in a multi-channel spectrometer vary widely. At the present time the instruments are constructed with transistor equipment, which has allowed a sensible reduction in size. The number of channels goes into several thousands, but most frequently is in the range of 200 and 400.

References

1. Price W. J., *Nuclear Radiation Detection*. McGraw-Hill, New York 1964.
2. Fassbender H., *Einführung in die Messtechnik der Kernstrahlung und die Anwendung der Radioisotope*. Thieme, Stuttgart 1962.
3. Snell A. H., (Ed.), *Nuclear Instruments and Their Uses* (Vol. I.). Wiley, New York 1962.
4. Fenyves E., Haiman O., *Die physikalischen Grundlagen der Kernstrahlungsmessungen*. Verlag der Ungarischen Akademie der Wissenschaften, Budapest 1965.
5. Rossi B. B., Staub H. H., *Ionization Chambers and Counters*. McGraw-Hill, New York 1949.
6. Sharpe J., *Nuclear Radiation Detectors* (ed. 2). Methuen-Wiley, New York 1964.
7. Wilkinson D. H., *Ionization Chambers and Counters*. Harvard Univ. Press, Cambridge 1950.
8. Taylor D., *The Measurement of Radioisotopes*. Wiley, New York 1958.
9. Watt D. E., Ramsden D., *High Sensitivity Counting Techniques*. MacMillan, London 1964.
10. Birks J. B., *The Theory and Practice of Scintillation Counting*. Pergamon Press, London 1964.
11. Curran S. C., *Luminescence and the Scintillation Counter*. Academic Press, New York 1953.

12. Dearnaley G., Northrop D. C., *Semiconductor Detectors for Nuclear Radiations.* Wiley, New York 1963.
13. Burge E. J., (Ed.), *Semiconductor Detector for Nuclear Radiations.* E. and F. N. Spon, Ltd., London 1966.
14. Siegbahn K., (Ed.), *Alpha, Beta and Gamma-Ray Spectroscopy.* Nort-Holland Publ. Co., Amsterdam 1964.
15. Erbe H., Atompraxis **10**, (7), 1 (1964).
16. Salmon L., *Gamma-Ray Spectroscopy Applied to Radioactivation Analysis* (Part 1). AERE C/R 2377 (1), Harwell 1959.
17. Crouthamel C. E., *Applied Gamma-Ray Spectrometry.* Pergamon Press, London 1960.
18. Rakovič M., *Activation Analysis.* Ilife, London; ČSAV, Prague 1970.
19. Siegbahn K., *Beta and Gamma-Ray Spectroscopy.* Interscience. New York 1955.
20. Slunečko J., Proc. Symp. Activation Analysis, Modra-Piesky (Czechoslovakia) 1963, 82.
21. Covell D. F., Anal. Chem. **31,** 1785 (1959).
22. Connally R. E., Leboeuf M. B., Anal. Chem. **25,** 1095 (1953).
23. Mc Isaac D. L., US Naval Radiological Defense Laboratory Tech., Rep. USNRDL-TR-72, 1956.
24. Geisler G., Brunner G., Atomkernenergie **7**, 2 (1962).
25. Anders O. V., Beamer W. H., Anal. Chem. **33,** 226 (1961).
26. Kukula F., Slunečko J., Šimková M., Report ÚJF, No. 679/1962.
27. *Radiografia*, Almanac, Izd. inostran. lit., Moscow 1952.
28. Yagoda H., *Radioactive Measurement with Nuclear Emulsions.* Wiley, New York 1949.
29. Boyd G. A., *Autoradiography in Biology and Medicine.* Academic Press, New York 1955.
30. 20th Century Electronics, Ltd., England: Boron Trifluoride and Boron Lined Neutron Detectors for Nuclear Instrumentation (Project Report).
31. Brousil J., Dienstbier Z., (Ed.), *Radioisotopové diagnostické metody (Radioisotope Diagnostic Methods).* SZN, Prague 1965.
32. Ljunggren K., *Radioisotopes in the Physical Sciences and Industry.* IAEA, Vienna 1962, p. 199.
33. Ljunggren K., Pure Appl. Chem. **1,** 31 (1960).
34. Borg D. C., Segel R. E., Campbell E., Int. J. Appl. Rad. Isotopes **11,** 10 (1961).
35. Adams F., Hoste J., Atomic Energy Rev. **4**, 113 (1966).

4. Characteristic Features of Radioactive Substances and Principles of Experimental Techniques

V. KRIVÁŇ

4.1 Introduction

When radioactive substances are to be used either as radioactive indicators, or as sealed sources of radiation, account must be taken of their many peculiar properties which are unknown in work with inactive materials. A thorough examination of each problem is necessary to decide whether a radiochemical method should be given priority over other methods to solve a particular analytical task. On the other hand, a knowledge of these peculiarities will often prove decisive in the choice of the particular radionuclide, method and experimental technique.

The use of radioactive materials as indicators in analytical chemistry is based on the fact that there is no essential difference in chemical properties of a radioactive and an inactive isotope of the same element. There are circumstances, however, under which radioactive elements behave differently from inactive ones.

Radioactive nuclides are for the most part prepared and very often employed in imponderable and invisible amounts. When such minute amounts of materials are involved, deviations from states of chemical equilibrium may ensue and account must be taken in particular of adsorption processes that occur at the boundary between phases. Considerable loss in radioactive material may occur due to adsorption on walls of the container, on filter paper or on precipitates. In addition the handling of minute amounts of radioactive elements is accompanied by a number of further difficulties, e.g., the formation of radiocolloids, or the impossibility of performing a rapid operation, like precipitation. Many of these inconveniences may be overcome by adding a carrier of the radioactive element under investigation. We generally designate as carriers inactive elements or their suitable compounds that are added to a radioactive substance in measurable amounts, which have identical or very closely related chemical properties. The use of a carrier likewise entails certain difficulties arising from a change of the specific activity of the radioactive substance.

Another feature of the methods involving radionuclides—in comparison with other methods—is their extraordinary sensitivity, which depends on the specific activity of the tracer compound employed. Maximum sensitivity is attained when radionuclides are used as indicators without carrier (carrier-free radioisotope). By measuring radioactivity we may thus determine extraordinarily small quantities of substances, of the order of 10^{-15} to 10^{-18} g. The extremely high sensitivity of these methods puts corresponding claims on the purity of the substances, not only as regards chemical purity, but above all, radionuclide and radiochemical purity (see 4.3.1). With such a high degree of sensitivity, even quite minute amounts of radionuclide impurities, which have no effect whatsoever on the chemical purity, may be detrimental and lead to faulty results and erroneous conclusions. Besides foreign radioactive elements, other compounds of the basic radioactive element, which, of course, have different chemical properties may adversely affect the experiment.

The application of radionuclides makes use of the properties of radioactive radiation. Hence, we must take into account the laws of radioactive transformations which determine the time course of radioactive decay. Often rapid manipulating procedures are required for work with radionuclides having a short half-life. The peculiar properties of radioactive substances affect their special quantification, special methods for radionuclide identification, and specific problems that may arise when various analytical operations are carried out, or samples are prepared for measurement.

A further significant peculiarity is that all work with radioactive substances involves some danger and requires that safety precautions be taken. Every type of radioactive radiation is hazardous, particularly at higher doses, to biological systems. One of the basic tasks in work with radioactive substances is to prevent internal contamination of the body, or its external irradiation. The safety rules by which this is attained impart a special character both to the experimental methods involved, and to the laboratories themselves.

The present chapter describes and discusses the most important characteristic features of radioactive substances and the employed experimental techniques. In the case of some more specific questions, whose detailed treatment falls beyond the scope of this book, the reader will at least find literary references to the subject.

4.2 Quantification of Radioactive Substances

In current chemical or physical practice the quantity of a substance is expressed either in weight or in volume unit. The amount of radioactive nuclides could essentially be likewise determined by their weight, but for the most part this is unnecessary. These amounts are so small that it is difficult to measure their weight experimentally, and moreover, such a quantitative expression fails totally to characterize radionuclides from the viewpoint of their function, which is always related to their spontaneous decay with simultaneous emission of radioactive radiation, i.e., to their radioactivity. Thus, it is far more important to express the amount of radionuclides in terms of their radioactivity (henceforth only "activity").

For the most part it is not necessary to know the absolute activity. Thus, for instance, when radioactive substances are applied in tracer techniques, only relative measurements are carried out. But there are cases when the absolute activity must be known. Therefore, in this section we shall briefly review terms, quantities and units that are of importance from the viewpoint of quantification of radioactive substances, and also of some methodical data important in practice.

4.2.1 Activity and its Units

Radionuclides are nearly always specified by means of their activity, which is characterized by the number of their atoms disintegrating per unit time. The activity expresses the potency of the radioactive source, and by inference, also the amount of the radioactive nuclide.

The basic unit of activity is the *curie* (Ci), which is defined by the I.C.R.U. as exactly 3.7×10^{10} disintegrations per second. The millicurie (mCi – 10^{-3} Ci), the microcurie (μCi – 10^{-6} Ci), the nanocurie (nCi – 10^{-9} Ci) and the picocurie (pCi – 10^{-12} Ci) are practical units also in common use.

A further unit, more rarely used, is the *rutherford* (rd), which is defined as 10^6 disintegrations per second.

In radiochemical work, the minute is frequently used as a time unit for the number of disintegrations, or of counts recorded by an instrument. The conversion factors for various activity units are given in Table 4.1.

Table 4.1

Conversion Factors for Various Activity Units

	d min^{-1}	d sec^{-1}	rd	mCi
1 d min^{-1}	1	1.67×10^{-2}	1.67×10^{-8}	4.5×10^{-10}
1 d sec^{-1}	60	1	10^{-6}	2.7×10^{-8}
1 rd	6×10^{7}	10^{6}	1	2.7×10^{-2}
1 mCi	2.22×10^{9}	3.7×10^{7}	37	1

Activity A may be expressed from the known quantity (number) of radioactive nuclei N and the decay constant λ, or the half-life T, as

$$A = -\frac{dN}{dt} = \lambda N = 0.693 \frac{N}{T}. \tag{4.1}$$

As seen from this relation, the activity of the sample is directly proportional to its radionuclide content. The activity of various preparations with an equal content of radioactive nuclides will differ according to the value of the decay constant (the half-life). The higher the rate of disintegration of the given radionuclide, the higher will be its activity.

This is further related to the fact that the activity is not constant in time, but decreases exponentially, at a rate that depends on the nature of the radionuclide. Activity at any other time may then be simply calculated by means of the exponential law (2.2).

When calculations are made frequently, reference may be had to current radionuclide tables which set down the values of the functions $e^{-\lambda t}$ and $1 - e^{-\lambda t}$. These permit the ratios of residual and decaying activities to be read off directly from the initial activity.

In rapid orienting calculations time is expressed in terms of half-life ($t = n T$) and the activity ratio is given as

$$\frac{A}{A_0} = 2^{-n}. \tag{4.2}$$

In the case of radionuclides with a long half-life, or in general, when the period considered is very small in comparison with the half-life period ($t \ll T$), the following approximated relations may be used in the calculations

$$\frac{A}{A_0} = 1 - \lambda t; \qquad \frac{A_0 - A}{A_0} = \lambda t. \tag{4.3}$$

From a practical point of view a quantitative specification of radioactive sources on the basis of their absolute activity is often suitable only in cases where self-absorption is without significance, as for instance, in sources of small mass and in gamma radiation. In the case of beta radiation sources, the number of beta particles emitted from the source does not correspond, for the most part, to the real number of disintegrations, but is lower, for part of the particles become absorbed within the sample. This effect may also occur with soft gamma radiation. For this reason, the "effective activity" is often of more interest. It may be expressed in various ways, but in every case the activity corresponding to radiation leaving the source must be stated, and not that corresponding to the real number of disintegrations.

Instead of the term "effective activity," we often meet, mainly in connection with gamma radiation sources, with that of "equivalent activity," which is defined as being equal to that of a point source of the same radionuclide emitting the same radiation intensity at an equal distance from the center of the source [1].

In the application of radionuclides we often make use simply of "detected activity." The mutual interrelation between absolute and "detected" activity is affected by numerous factors which depend on particular experimental conditions (see also Chapter 3).

4.2.2 Specific Activity

The material used for tracer work rarely, if ever, consists of the pure radioisotope. The proportion of the active species is frequently expressed in terms of specific activity of the material, defined as the ratio of the number of atoms of the given radioactive isotope to the total number of atoms of the same element. The specific activity, as defined in this manner, is thus the fraction of the given element that is present in the form of the particular radioisotope.

For practical purposes, specific activity is expressed in terms of the disintegration rate per unit weight or unit volume, e.g., $mCi\,g^{-1}$, $mCi\,ml^{-1}$, $mCi\,millimole^{-1}$, etc.

The estimation of absolute specific activity is based on the determination of the number of disintegrating nuclei of the radionuclide per unit time, and is similar to the absolute determination of activity in general, as described under 4.3.

For current work with radioactive substances it is generally sufficient to know the relative specific activity. In counting methods this is expressed by the number of recorded pulses, and in other methods, by the value of various data (e.g., the number of scale divisions in the case of an electrometer), per unit time, per unit volume, weight or per millimole. The entire series of samples whose relative specific activity is mutually compared must be measured on the same instrument and under identical conditions. The simplest to determine is the specific activity of solutions. A portion of the solution is pipetted out and is then processed by one of the procedures for sample preparation, as described later (Chapt. 4.8). If the radioactivity of the solution is too high to measure with sufficient accuracy, the original sample should be first diluted.

The measurement of the specific activity of solids is generally more difficult. The radioactive preparation may be processed in solid form; nevertheless, it is more convenient to dissolve it and process it as a radioactive solution.

Certain experiments involving a strong dilution of the radioactive element require the initial specific activity to be sufficiently high. If the stated or determined radioactivity is lower than that required, concentration of the radioactive element must first be increased by means of radiochemical operations. The concentration of diluted solutions can be relatively easily performed by separation of the radioactive element with an isotope or an inert carrier. However, to increase the concentration of a radioactive element is sometimes highly complicated. An essentially simpler case is when the specific activity is higher than that which is required. A decrease of the specific activity is achieved by a corresponding dilution of the preparation by means of a solvent in the case of solutions, and by separating radioactive isotope with the required amount of an isotopic or an inert carrier in the case of solid substances.

4.2.3 Relation between Weight and Activity of Radionuclides

Work with radioactive substances involves for the most part such small amounts that they cannot be weighed on an analytical microbalance (imponderable amounts). Yet, there are cases where the weight of the sample is of interest. This is especially true for radionuclides with a long half-life, where we currently meet with amounts that are of interest not only in the operational scale of microtechnique, but sometimes also in that of normal laboratory technique. Thus, for instance, 1 mCi of radionuclide ^{14}C weights 0.224×10^{-3} g.

If the sample contains m grams of a radioactive nuclide whose atomic weight is M, then the number of radioactive atoms may be expressed as

$$N = \frac{m \, . \, N_0}{M}, \qquad (4.4)$$

where N_0 is Avogadro's number (6.023×10^{23}). From relations (4.1) and (4.4) we obtain for the weight of the radioactive nuclide

$$m = 2.40 \times 10^{-24} \, AMT, \qquad (4.5)$$

where A is the number of disintegrations per whatever time unit in which T is expressed. If we express the activity in mCi and the half-life in days, then we can obtain the weight of the radionuclide in g from the relation

$$m = 7.65 \times 10^{-12} AMT. \tag{4.6}$$

4.2.4 Dilution of Radioactive Solutions

Often radioactive solutions have to be diluted. At the same time it is necessary for the most part to calculate what volume V (in ml) of the original concentrated radioactive solution of absolute specific activity a (mCi ml^{-1}) must be removed to obtain volume V_1 of the diluted solution with a relative specific activity I_1 (cpm ml^{-1}). The relative activity of the whole volume V_1 is given by

$$I = I_1 V_1 \text{ cpm} = \frac{I_1 . V_1}{60} \text{ cps.} \tag{4.7}$$

The number of disintegrations per second (d sec^{-1}) in this volume equals

$$a_1 = \frac{I_1 . V_1}{60\,\eta}, \tag{4.8}$$

where η is the proportionality factor between the measured and the absolute sample activity (efficiency coefficient). To prepare a diluted solution with a total absolute activity a_1, we need V ml of the original solution

$$a_1 = a \times 3.70 \times 10^7 \times V \text{ d sec}^{-1}. \tag{4.9}$$

From a comparison of the two preceding equations we obtain the relation

$$\frac{I_1 . V_1}{60\,\eta} = a \times 3.7 \times 10^7 \times V \tag{4.10}$$

by means of which we can easily express the dilution ratio

$$\frac{V_1}{V} = \frac{a \times 3.70 \times 10^7 \times 60\,\eta}{I_1}, \tag{4.11}$$

and eventually the volume V which must be removed from the original concentrated solution.

4.3 The Concept of Purity and its Estimation

4.3.1 Definition of Purity

In current chemical practice there exists a single criterion for the purity of a substance. In the case of radioactive materials, purity must be essentially judged in a complex manner from three aspects, viz., chemical, radionuclidic and radiochemical.

Chemical purity. We may consider a radioactive sample as chemically pure when neither physical nor chemical methods can prove the presence of other interfering materials. However, it may contain imponderable traces of other radioactive elements that are unweighable and undetectable by relevant analytical methods, but that manifest themselves by their characteristic, and eventually very high activity.

Radionuclide (radioactive) purity may be defined as the ratio of activity of a given radionuclide to the total radioactivity of the sample. A preparation is pure in this aspect when it contains a single radionuclide in a measurable quantity, though it may contain a large amount of stable nuclides. If through the decay of the radionuclide, its daughter radionuclide is formed, the latter is not usually considered as a radionuclide impurity. Thus, for example, radionuclide ^{90}Sr will always contain its daughter radionuclide ^{90}Y. The part of the total activity corresponding to the daughter radionuclide will depend on the half-life of the parent and the daughter radionuclide. In every case, however, the presence of radioactive daughter products have to be taken into consideration.

Radiochemical purity is defined as the ratio of activity of an element in a given chemical form to the total activity. That substance is considered as radiochemically pure which contains the radionuclide exclusively in a single chemical form.

4.3.2 Sources of Impurities

Radionuclide impurities may arise from several sources. One of them is represented by impurities of the target material employed in the preparation of radionuclides. Further such sources may be other elements present in such amounts that would be insignificant even in spectrally pure material but which in certain cases provide, by irradiation, fairly marked activity in comparison with the required radionuclide. This occurs when the activation cross-section of a trace impurity considerably exceeds that of the matrix element. An example of such target material is, in neutron activation nickel containing traces of cobalt.

Another significant source of radionuclide impurities is provided by nuclear side reactions taking place during activation. The elements present and used as target material are, for the most part, polyisotopic. From the aspect of the desired nuclear reaction, they contain, besides the matrix nucleus, one or more further types of nuclei which, under the impact of the same bombarding particles, are subject to other types of nuclear reactions, the results of which are, of course, other types of radionuclides. In some cases, undesirable side reactions may take place even with one and the same nuclide.

The content of radionuclide impurities arising from these two sources may be controlled by careful choice of irradiation conditions and a suitable subsequent processing of the irradiated target.

Radiochemical impurities in labeled compounds have their sources in the conditions of preparation and storage. For example, during a synthesis of labeled compounds, a whole series of other tracer compounds, besides the main product, may be formed in various amounts. Similarly, radiochemical impurities may originate also with other methods of preparation of tracer compounds. In case of direct activation this occurs as a result of the Szillard-Chalmers effect, and in the preparation by exchange reactions it is due to competitive exchange reactions. Labeled compounds are usually prepared in small quantities, which prevents purification by current methods, such as distillation, crystallization, etc. On the other hand, chemical changes may ensue during storage as a result of the high specific activity of the labeled compounds. Such changes giving rise to radiochemical impurities are brought about mainly by the radiation effects of radiation emitted by the radionuclides [2–11], but also by chemical decomposition which is not connected with radiation. Today several ways of minimizing the formation of radiochemical impurities due to radiation chemical effects, are known for several organic materials.

Every decomposition of radioactive substances is dependent on time and therefore, longer storage should be avoided. The rate of decomposition is also markedly affected by the specific activity: the higher the specific activity, the more favorable are the conditions for the formation of radiochemical impurities through radiation decomposition. Thus the need for a lower specific activity to reduce decomposition and that for a higher one for experimental reasons oppose each other. However, we should, if possible, at all times avoid high specific activities.

Radiochemical impurities during storage may also ensue through an exchange reaction between a radioactive isotope in the labeled compound and a stable isotope of the same or some other compound. In order to be able to judge the importance of this source of errors, often present in an experiment, we must know the possibilities and conditions of the course of exchange reactions in a given system.

In general, the rate of contamination depends to a considerable degree on the condition of storage. It may be reduced by storing the compound in suitable solutions or mixtures with an inert material, e.g., by adsorption on paper. In many cases it will be found useful to store it in a refrigerator.

Some useful information on sample-mounting and storage of labeled compounds may be found in the relevant literature [13].

4.3.3 Purity Requirements

In the case of radioactive materials these requirements vary considerably and depend mainly on the nature of the sample and its impurities, the manner of application of the radioactive materials, and the method selected for detection.

When radionuclides are used as radiation sources for methods based on radiation absorption and scattering, there is no claim for radiochemical purity and requirements for radionuclide purity are not very high.

Radionuclide and radiochemical purity are very important, when radioactive substances are used as indicators. Here, even minute impurities may cause considerable error in the analysis. Nevertheless, requirements for purity here, too, may vary considerably from case to case. High radionuclide purity is, however, one of the basic requirements. When the relative activity of the original indicator substance is measured, radionuclide impurities due to radioisotopes of the same element need not necessarily have an interfering effect, for they behave in the same manner as the applied radioisotope. The isotopic ratio in such a mixture does not change. However, the differences in half-lives have to be taken into account. An example for this case are the two isotopes of iron, ^{55}Fe (T = 2.9 years) and ^{59}Fe (T = 45 days). Here, and in similar cases, it will suffice to take into account the presence of the second isotope in the course of measurement and in the interpretation of the results obtained.

If the radionuclide impurity is due to a radioactive isotope of a foreign element and the latter will behave, in the course of the experiment, similarly as the element studied, then its presence is always undesirable. The range of tolerance, however, depends on the application concerned. Traces of impurities may affect the results considerably when the percentage of the radioactive substance involved in the given process is very small and when the process takes place in the presence of a small amount of impurity. This is the case, for instance, when the solubility (solubility product) of very slightly soluble substances is determined.

Exact data on the nature and amount of radionuclide impurities are of basic significance. In many cases such data enable a choice of working procedures will eliminate the effect of impurities on the results.

Requirements for radiochemical purity of labeled compounds will likewise differ according to the mode of application of the radioactive chemicals. In this connection it is always important to consider the excepted behavior of the radionuclide in the course of the proposed experiment. The radiochemical purity may be inadequate even though the radioactive compound contains only the right radionuclide present in the right type of molecule, but in its various positions. If this molecule suffers no change during the proposed experiment, then the site of marking is irrelevant.

In general, chemical purity of radioactive preparations is usually of less interest than their radionuclide or radiochemical purity.

4.3.4 Control of Radionuclide and Radiochemical Purity

Physical methods, generally employed for radionuclide identification (Chapt. 4.5), based on the determination of half-life, type and energy of the emitted radiation, will often prove adequate for a control of radionuclide purity.

Among physical methods for radionuclide purity control, gamma spectrometry occupies the foremost place. Alpha spectrometry provides data on alpha particle energies, and is practically limited to the identification of heavy elements. Beta spectrometry has so far been relatively little used in current analyses of radioactive material. The introduction of liquid scintillation spectrometers, however, has markedly enhanced its significance for determination of beta radiation sources in mixtures. Because of ist simplicity, the determination of maximum energy by the absorption filter method is widely used with beta radiation sources, but it is not sufficiently sensitive when the presence of soft beta radiation sources is to be proved. The determination of the half-life is of considerable importance, but may be applied to estimate radionuclide impurities only in relatively short-lived radionuclides.

The exclusive use of physical methods will often prove insufficient for radionuclide purity control.

The presence of a foreign radionuclide may be proved with high sensitivity by means of chemical separations. This method fails, of course, when the sample is contaminated with radioisotopes of the same element. In such a case data on admixtures can reliably be obtained solely by means of physical identification methods. In general, the best results in radionuclide impurity determination are obtained by a combination of chemical and physical methods—and above all, by the use of gamma spectrometry.

Once radionuclide impurities have been proved and identified, efforts are made to carry out the measurement in such a way that their presence need not be taken into consideration at all, or their contribution to the total activity be known and the final results corrected accordingly. It is a simple case when the admixed radionuclide has a less penetrating type of radiation than the radioactive indicator, or a substantially lower energy with the same type of radiation, since the lower energies can be easily eliminated in the detection process. When this fails, radionuclide impurities must be separated radiochemically. Radionuclide impurities are for the most part eliminated by the same methods as are used for their chemical proof. This purification must naturally be performed before the radioactive preparation is added to the sample. And of course, it is best to prevent the radioactive sample from becoming contaminated at all. In consequence, we shall always use substances of a high initial purity. In the preparation of radionuclides, those nuclear reactions are selected in which the probability of competing nuclear reactions, and thereby also of the formation of other radionuclides, is as small as possible.

Labeled compounds require a check for radiochemical purity. Methods such as those employed for determining the melting point, absorption spectra, etc., which are otherwise currently used, are unsuitable for this purpose. The method of radioisotope dilution analysis and chromatographic separation procedures may be used, however, with advantage in the control of radiochemical purity.

Reversed isotope dilution analysis may be conveniently employed to determine the purity of radioactive substances by dilution with an inactive carrier. This method may serve as the main component for the purity assay of the radioactive indicator, and also of the amount of specific impurities. In this procedure an excess quantity of the pure inactive substance is added to the radioactive sample. After thorough mixing, a portion of this mixture is stored, and the rest is purified by any suitable method. When a pure product is obtained, the specific activities of the purified and the original samples are determined and compared. If the original sample is pure, the two specific activities are identical. If, however, the specific activity following purification represents only a part of that of the original mixture, the sample is contaminated. The method of isotope dilution may fail in the case that the carrier is not pure, or the isotope effect is more marked, or the purification is incomplete.

Both paper chromatography and thin layer chromatography are widely used in radiochemical analysis. Suitable elution systems permit even very complicated mixtures to be separated. Paper or thin-layer chromatograms may be evaluated with the aid of a radiation detector, and on this basis we may determine the quantity of impurities present.

If a radioactive atom can possess different positions in a particular molecule and if only one of the positions is to be labeled, the above methods of radiochemical purity control will not suffice. In such a case, the purity control should be performed by means of a suitably selected chemical reaction (decomposition, hydrolysis, etc.) and a subsequent radiochemical analysis of the reaction products. The chemical analysis of radioactive samples with high specific activity are performed most frequently by means of emission spectrography, polarography and colorimetric absorption methods. The analysis of samples with low specific activity is carried out by means of other analytical methods.

Aspects of a practical interest in connection with purity of radioactive substances may be found in special publications [22].

4.4 Behavior of Radioactive Substances in Trace Amounts

4.4.1 Introduction

Chemistry deals with the study of chemical properties and conversions of substances in such amounts and concentrations as are readily accessible to chemical methods of identification and determination. The amounts that are usually employed in the investigation are greater than 10^{-6} g and the concentrations higher than 10^{-5} M. The validity of the laws describing chemical equilibria and kinetic phenomena refers to quantities and concentrations above the given limits. Making the concentration region below this limit accessible is the object of ultra-microchemistry, via physicochemical and physical methods.

A characteristic feature of work with radioactive materials is that they may be prepared, detected and applied in amounts and concentrations far below the limits perceivable by standard physical and chemical methods. These amounts of radioactive materials can neither be observed nor weighed. The behavior of radioactive substances in extremely small amounts and low concentrations differs significantly from their behavior when present in macroscale. The causes for this do not really lie in radioactivity and the emitted ionizing radiation. Under proper conditions even inactive materials in these quantities would show deviations from their behavior as displayed in macroscale. A simple detection of radioactive substances enables these peculiar phenomena to be observed. They are undetectable with other methods, and thereby are without any significance. The weights of carrier-free radioactive elements of 1 nCi activity, which can be measured relatively easily, and the corresponding concentrations when this quantity is dissolved in 1 ml, are given in Table 4.2.

In practice, however, these calculated quantities are obtained only in some cases, for radioactive preparations also contain a certain amount of stable isotopes of the same element which originates from target material or competitive nuclear reactions, or is introduced in the processing of target material following irradiation. In spite of this, in many cases smaller quantities and concentrations are involved in current chemical operations and they represent, in a certain sense, a peculiar state of the substance, manifesting itself by characteristic properties and effects.

Table 4.2

Weight, Specific Activity and Concentrations of Various Carrier-Free Radionuclides

Radionuclide	Half-life	Amount by weight at 1 nCi (g)	Specific activity (mCi/g)	Molarity of containing solution 1 nCi/ml
^{238}U	4.5×10^{9} yr	3.00×10^{-3}	3.32×10^{-4}	1.26×10^{-2}
^{36}Cl	3×10^{5} yr	3.03×10^{-8}	33.0	8.42×10^{-7}
^{14}C	5,760 yr	2.26×10^{-10}	4.42×10^{3}	1.61×10^{-8}
^{3}H	12.26 yr	1.02×10^{-13}	9.75×10^{6}	3.40×10^{-11}
^{35}S	87 days	2.41×10^{-14}	4.15×10^{7}	6.88×10^{-13}
^{32}P	14.2 days	3.47×10^{-15}	2.88×10^{8}	1.08×10^{-13}
^{131}I	8.04 days	8.07×10^{-15}	1.24×10^{8}	6.16×10^{-14}
^{24}Na	15 hr	1.15×10^{-16}	8.72×10^{9}	4.65×10^{-15}
^{132}I	2.3 hr	9.66×10^{-17}	1.03×10^{10}	7.32×10^{-16}
ThC″(^{208}Tl)	3.1 min	3.44×10^{-18}	2.91×10^{11}	1.65×10^{-17}

4.4.2 Indicator Concentrations and Validity of Basic Physical-Chemical Laws

A basic question in work with radioactive materials is whether the basic physical-chemical laws governing their behavior in macro-quantities and macro-concentrations may be applied to substances in very small quantities and concentrations, or whether their validity here is partly or entirely lost. This question is of considerable practical and theoretical significance for the separation and purification of radionuclides, and also for their applicability.

A statistical exposition of the laws governing chemical equilibrium leads to the conclusion that in systems comprising an enormous number of reacting particles, chemical states of equilibrium are not affected by the dimensions of the reacting system, i.e., by the volume and quantity, but solely by temperature and pressure or concentration. The energy distribution of reacting particles is described by the Maxwell–Boltzman statistics. When the number of reacting particles is reduced under a certain limit the statistical laws are no longer valid and the state of equilibrium becomes indefinable and irreproducible [14, 15]. Amounts that are close to these limit values are frequently applied in radiochemical studies. In reality, difficulties may be encountered with quantities of the substances that are several orders higher than those generally considered as extreme, as far as the validity of the statistical laws is concerned. However, the cause of these phenomena may be of a kinetic nature, or may reside in secondary and side processes, or even may be due to trace amounts of other substances simultaneously present in the system.

Equilibrium constants may, in the same system, acquire values differing by several orders, according to whether the reacting components have comparable concentrations, or whether one of them occurs in trace amount while the other is present in macro-concentration. This may lead to reactions which are entirely different from those occurring under ordinary conditions. Some examples are provided by Haissinsky [16].

Extremely low concentrations also cause a decrease in reaction rates due to the highly reduced number of effective collisions between the reactants.

4.4.3 Adsorption Phenomena

In extremely diluted solutions surface effects may considerably affect their behavior. Adsorption phenomena often leading to quite unexpected results take place at the liquid–solid interphase. The original amount of a substance in solutions (below 10^{-5} M) may substantially decrease as a result of adsorption and, in extreme cases, even totally disappear.

Adsorption phenomena encountered during manipulations with radioactive substances may be of interest for two reasons: on the one hand, they lead to undesirable losses during a chemical (radiochemical) operation; on the other hand, adsorption may be an effective method for the separation of radionuclides.

Adsorption on Glass and Other Materials

Adsorption occurs, before all else, on the walls of the vessels and on various materials with which the solution may come into contact, e.g., filter paper. Hence it becomes imperative in radioanalytical practice to know the degree of adsorption.

Glass containers are suitable for the storage of low-concentration, and above all of non-aqueous radioactive solutions containing organic compounds. However, reactions between the organic compound and alkalis released from glass may sometimes occur.

Solutions containing inorganic substances should not be stored in glass bottles and especially not the aqueous solutions at high pH. Thus, for example, radiosodium is readily and irreversibly adsorbed on glass; a high degree of adsorption occurs also in the case of silver ions, phosphates and other radioactive elements in the form of radiocolloids.

The adsorption of cations on glass is explained by the exchange on groups such as $\geqslant$SiOH and $\geqslant$SiOX (X represents a metal cation) present in the glass. The adsorption of ions present in diluted solutions on glass is pH dependent (e.g., Tl^{+} adsorption increases with increasing pH, Zr^{4+} adsorption has a maximum value [19]). In strongly acid solutions the adsorption on glass is insignificant. The presence of additional electrolytes in the solution and also the structure of the glass surface have considerable effect on the adsorption [17, 18]. The adsorption ability of glass is enhanced by a preceding action of lyes which disturb the hydrated surface layer. This gives rise to a dissociation in an aqueous medium of the groups $\geqslant$SiOX into SiO^{-} and X^{+} far more easily than of the group $\geqslant$SiOH. Thus, for instance, the amount of adsorbed radionatrium has been observed to increase with preceding warming-up of the glass over a flame [17].

If solutions of radioactive substances are introduced into glass vessels for evaporation, the adsorption causes great difficulties. The adsorbed substance adheres firmly to the walls and it is extremely difficult to recover it quantitatively again. In many cases it was found suitable to apply a silicon-coating wax or resin to the walls, thus making them inaccessible directly to the solutions.

For the above reasons polyethylene and other plastics are used in the radiochemical experiments. However, at high activities their surface may become affected by the radiation. In general, plastic materials are not suitable for diluted solutions of organic substances. Sometimes other materials, mainly metals such as platinum, gold, nickel, etc., are employed for specific purposes.

Cellulose possesses also ion-adsorption properties. Carboxyl groups are considered to be active in the ion exchange processes. The affinity for cations of cellulose is the same as that described for ion exchange resins containing carboxyl groups. Anions are adsorbed on cellulose in a substantially lower degree and the mechanism of this event has not been satisfactorily elucidated as yet. In some cases, adsorption may take place in the colloid form of the particles which are fixed in the pores of filters.

Adsorption should be taken into account at the outset, if filter paper is used. A disregard for adsorption could result in considerable error. In some cases as much as 98.5 % of the total radioactive element (radium sulphate) was adsorbed on the filter paper [20, 21]. In paper chromatography also, the R_F values remain independent of the respective quantities of the elements only insofar as there is no disturbing effect due to adsorption.

Adsorption of radioactive elements on crystals and precipitates plays an important role in radiochemical analytical practice. Here the microcomponent is either adsorbed on the crystal surface (external adsorption), or migrates into the crystal lattice (internal adsorption).

The external adsorption is attributed to the existence of a electrical double layer at the dividing line between crystal and solution. An efficient adsorption of a trace element occurs when its charge is opposite to that of the crystal. The latter is determined by the nature of the ions present in excess amount in the solution. Thus, for example, the amount of ThB^{2+} adsorbed on AgBr increases with an increasing bromide ion concentration in the solution. Similarly, the amount of Ra^{2+} adsorbed on calcium sulphate precipitate increases by increasing the amount of excess sulphuric acid. An excess of calcium chloride in the solution, which supplies positive charges to the precipitate, causes a decrease in the degree of the adsorption [16].

The efficiency of removing a microcomponent from the solution by adsorption is affected also by other factors: the surface area of the precipitate, the solubility of the compound of the trace ion and the macrocomponent ion, the valence of the microcomponent, the presence of foreign ions, and aging of the precipitate. The larger the surface of the precipitate, the more tracer element is adsorbed; the adsorption increases with decreasing solubility of the compound formed from the trace ion and the ion of the macrocomponent with an opposite charge in the solution; the higher the charge of a trace ion, the more easily the latter is adsorbed; the presence of foreign, primarily multivalent, ions with the same charge as the radioactive trace ion, reduces the adsorption of the latter. Aging of a precipitate may result in a decrease or an increase of adsorption, depending on the accompanying phenomena.

In the case of the adsorption of non-electrolytes at the crystal surface the phenomenon is due to the action of van der Waal's forces.

A characteristic feature of these types of adsorption is that the adsorbed ions do not reach the crystal lattice but only surround the charged crystal surface.

However, there exists yet another mode of adsorption, based on an exchange between ions at the crystal surface and ions in the solution. Thus some of the ions from the solution reach the upper layers of the crystal lattice. An example of this type of adsorption is the replacement of surface barium ions of a barium sulphate precipitate by calcium ions in the solution. Calcium sulphate and barium sulphate are not isomorphous. The replacement occurs more readily when isomorphous compounds are involved.

In the case of internal adsorption the microcomponent is incorporated into the crystal lattice. Of all the adsorption processes, that of internal adsorption has so far been least elucidated. The transition of a trace element from the solution into the crystalline phase is probably promoted by active centers (higher energy sites) present in every real crystal [24]. There exists a large variety of more or less divergent views on this matter. One of these emphasizes the prominent role of vacancies in the adsorption [25].

References [15, 16, 19, 23–31] give clues to abundant and versatile material dealing with adsorption of radioactive elements on solids.

4.4.4 Precipitation and Crystallization

Separation by precipitation is one of the most commonly used classical methods of analytical chemistry. In most cases, particularly at trace levels, in which the amount of radioactive material is small, the concentration in solution before precipitation is too low to permit precipitation by exceeding the solubility of a compound considered to be "insoluble."

In current analytical practice we consider as practically insoluble, e.g., every binary compound whose solubility product is less than 10^{-10}. The concentration of the compound in the saturated solution is of the order of 10^{-5} M. Tracer concentration of an element can be as low as 10^{-12}–10^{-16} M. Often, even when the solubility product is exceeded, the amount of the precipitate thus formed is so small that for the most part only a colloid solution or an unfiltrable turbidity is produced.

A relatively simple case ensues when a given radioactive element is not required to remain in a carrier-free state. Here, to the radioactive element is added its inactive isotope in a weighable amount as carrier. Further processes are performed according to current methods. Carriers will be discussed in more detail in Chapt. 4.4.8.

The conditions become somewhat more complicated when the radioactive element is to be recovered from the solution without a carrier, or with only a negligable amount of it. In such a case we apply a non-isotopic carrier and remove the radioactive element by precipitation or crystallization.

The most important removal process is the formation of isomorphous crystals during crystallization or precipitation. When the radioactive element is isomorphous with the macrocomponent, it is incorporated through isomorphous replacement into the crystal lattice. Co-precipitation of radium with barium in the form of sulphates, may serve as an illustration. The characteristic features of an isomorphous co-precipitation may be summarized as follows: The distribution of the trace element in the precipitate is, for the most part, homogeneous. The degree of co-precipitation is very little affected by conditions of crystallization and precipitation such as temperature, pH value, rate of separation, size of crystals, excess of the precipitant, etc. The microcomponent is firmly bound in the precipitate and cannot be removed from it by washing.

The qualitative rules governing co-precipitation and co-crystallization have been formulated by Paneth [32, 33] and Fajans [34, 35]. Credit for a systematic study of co-precipitation and co-crystallization and for an elucidation of the quantitative aspects goes to Hahn's [24, 36, 37] and Khlopin's [38–40] schools.

For true co-precipitation of the isomorphous–replacement type the progress of crystal separation and the detailed distribution of the active traces within the precipitate crystals may tend to approach either of two limiting laws. If the entire crystal is brought into equilibrium with solution, the trace element distribution is homogeneous and we obtain the *Berthelot–Nernst homogeneous distribution law* [68] expressed by

$$\frac{x}{a - x} = \frac{D \, . \, y}{b - y}, \tag{4.12}$$

where x and y are the amounts of micro- and macrocomponent respectively present in the crystalline phase, a and b the initial concentrations of the respective components in the solution, and D the enrichment (crystallization) factor.

Under certain circumstances when there is no equilibrium, the distribution of the trace element between the crystallic phase and the solution is described more correctly by *Doerner–Hoskins' logarithmic relation* [41]

$$\log \frac{a}{a - x} = \lambda \log \frac{b}{b - y}, \tag{4.13}$$

where x, y, a and b have the same meaning as under (4.12), and λ is the logarithmic crystallization constant.

Co-precipitation from supersaturated solutions and co-precipitations in which finely divided precipitate is allowed to stand in contact with the solution are likely to follow the homogeneous distribution law. The logarithmic distribution law is closely approached in co-precipitations produced by gradual evaporation with care to avoid any supersaturation, and for co-precipitations from supersaturated solutions that are vigorously stirred and quickly filtered.

Under special circumstances, radioactive elements may be separated in a carrier-free state from a solution even if the solubility product of the given compound has not been reached (see the next chapter).

4.4.5 Radiocolloids

Solutions of radioactive substances present in trace amounts often acquire the character of colloid solutions. This is the behavior of the majority of substances, which in an inactive form and in macroquantities would form precipitates. Colloid solutions of radioactive agglomerates, present in unweighable and invisible amounts, are termed *radiocolloids*. They may be produced at concentrations lower than or in excess of those corresponding to the solubility product.

The causes of the formation of radiocolloids have not been as yet sufficiently elucidated. According to prevalent interpretations radiocolloids may exist as *true colloids* [42, 43] and as so-called *pseudo-colloids* [26, 44, 45]. True radiocolloids are made up solely of a radioactive trace element and their origin is due to the low solubility of the corresponding compounds of the radioactive elements. Pseudo-colloids are inactive ultramicro-particles, or colloid particles of other substances that may be present even incidentally in the solution, as for instance, dust, hydrated SiO_2, hydrolysis products, etc., and these accumulate the radioactive ions or molecules.

The role of particular processes may be altered substantially, depending on the concentration of the radioactive element, the nature of the solution, and other conditions. Pseudocolloids may be formed also in extremely diluted solutions.

Due to radiocolloid formation, solutions of certain radioactive ions show unusual properties which are difficult to control. These are, among others, inhomogeneous distribution of a radioactive substance in the solution, its retainment on a filter, decreased diffusion rate, failure of dialysis through permeable membranes, coagulation or peptidization by electrolytes, adsorption (generally irreversible) on finely divided particles or on the walls of the vessel, anomal direction of electrolysis, nonspecific adsorption on an ion exchanger, etc. Ignorance of them might result in serious errors. In order to rule out these pecularities in the behavior of solutions of radioactive elements, the colloid particles must be either totally removed, or reduced to a minimum. On the other hand, however, radiocolloids may be advantageously utilized to separate carrier-free radioactive nuclides. To this end, use is made chiefly of the powerful radiocolloid adsorption.

The production of radiocolloids is affected in an essential degree by external conditions. The tendency to radiocolloid formation increases as the compound that comprises the radioactive element becomes less soluble. Consequently, the formation of radiocolloids is promoted either by the presence of an inactive macrocomponent in the solution, which leads to the formation of an insoluble component, or by the hydrolysis of the radioactive element. On the other hand, radiocolloid formation is inhibited by substances that form complexes with the radioactive element. The presence of foreign particles favors the formation of radiocolloids. The size and sign of the electric charge of a radiocolloid micelle (composition of the electrical double-layer) may be affected by the presence of foreign ions in the solution. The formation of radiocolloids may further be influenced by the nature of the solvent and by the presence of an inactive isotope. The capacity for radiocolloid formation has been observed so far with compounds of a large number of elements, e.g., Be, Mg, Sc, Ti, Mn, Fe, Co, Y, Zr, Nb, Ag, Sn, La, Ce, Au, Th, Pa, Pu.

Radiocolloids are studied by ultrafiltration, dialysis, diffusion, sedimentation, ion exchange adsorption, and electrolysis, and also by autoradiography and electrophoresis [16, 19, 42, 46].

Radioactive atoms may accumulate also in the gaseous phase on solid floating microparticles, forming the so-called *radioaerosols*. The essence of the aerosol resides in the inactive particle, on the surface of which the radioactive substance is adsorbed or condensed. Aerosol formation is considerably affected by the behavior of the radioactive element in the gaseous state.

4.4.6 Distribution in Liquid Systems

A primary point of interest, in view of the great practical significance of extraction methods, is the distribution of radioactive elements between immiscible solvents.

The distribution of substances between two liquid phases is given by the distribution coefficient α, which according to *Nernst's law* is defined as the ratio of equilibrium concentrations of a given component in both phases ($\alpha = c_1/c_2$). This ratio is independant of concentration at a constant temperature if neither dissociation nor association of the molecules occurs in one, or in both phases, i.e. the molecular state of the compound remains unchanged.

The validity of Nernst's law may also be extended to extremely low concentrations if the above conditions are fulfilled, as has been proved experimentally [47, 48]. Deviations from the distribution law may occur, however, in several cases. This is due, in the first place, to the formation of radiocolloids which may preferentially concentrate at the phase boundaries or on the walls of the vessel. A further cause may reside in adsorption on the walls of the vessel and in dissociation or association equilibria. These interfering factors should be taken into account in the distribution of trace amounts of radioactive elements, together with another fact, viz., that the activity of each phase gives the total concentration of the radioactive atoms regardless of their chemical forms.

The application of liquid-liquid extraction is dealt with in Chapt. 4.7.7.

4.4.7 Additional Special Properties and Effects

Radioactive substances present on the surface of another material in microquantities behave during evaporation differently from similar inactive substances present under the same conditions in macroquantities. Volatility of a radioactive substance does not depend solely on its properties, but also on the nature of fixation on the surface, on the type and properties of the surface and the composition of the ambient atmosphere. Similarly, deviations may also be observed during evaporation of radioactive substances.

The validity of Nernst's law does not hold in electrochemical processes when the concentrations are so low that the atoms of the radioactive element fail to cover the whole electrode surface with at least a monoatomic layer. In such a case the condition fails to be fulfilled that the electrolytic dissolution pressure of the metal is constant and that it does not depend on the nature of the electrode. This occurs fairly often in radiochemical practice if we consider that 5–10 ml of 10^{-7}–10^{-6} M of solution are required to produce a formal monoatomic layer on an electrode of 1–2 cm^2 in area after complete deposition of the ions from the solution.

As a result of extremely low concentrations, peculiar effects may also come into play in further chemical or physical processes.

4.4.8 Carriers

Many of the inconveniences referred to above may be overcome when a certain inactive substance is added in macroscopic amount to the trace radioactive element. This substance, generally designated as *carrier*, has from the viewpoint of the required processes and methods, properties either identical or closely related to the radioactive element. In a chemical sense, the carrier substance takes on the function of the radioactive element and *carries* it in the course of the chemical operations. Thus we can handle a larger amount of substance, which follows the well-known laws of physical chemistry. This is true when the radioactive substance and carrier is thoroughly mixed.

The mutual relationship between the carrier as the macrocomponent and the radioactive substance as the microcomponent may assume different characters. Accordingly, we distinguish three types of carriers.

The simplest situation is when the radioactive element need not be recovered without the carrier. Here, the so-called *isotopic carrier*, i.e., the inactive isotope of the given element, may be used. Nevertheless, it will fulfill the carrier function only when both substances are in the same chemical form. The chemical form of a radioactive element may generally be presupposed from the preceding chemical processes. If, however, the chemical form is not definitely established, the best approach is to add the carrier element in all the chemical forms that should be taken into consideration. The mixture thus obtained is then subjected to such chemical operations as will assure the transformation of all these various forms into a single one. Certain differences in chemical behavior may ensue in virtue of the isotope effect. They are, however, generally negligible even with the lightest elements. The isotopic carrier cannot be separated from the radioactive substance by any chemical method. Therefore, this type of carrier may be employed only in cases where even a lower specific activity will suffice for further operations.

If, however, the radioactive substance will subsequently have to be recovered without its carrier, or with only a negligible amount, then it is more expedient to employ *non-isotopic* (*inert*) *carriers* which may be divided into two groups, according to their relation towards the radioactive element.

The relation between the carrier and the tracer substance may be one of chemical similarity in view of the given reaction or process. Under a different reaction their behavior may vary. Thus, reactions and processes in which both components behave differently permit them to be again separated, if required. The function of a carrier of this type is most frequently fulfilled by elements of the same subgroup in the periodic table as the given radioactive element, e.g., La for Ac, Ba for Ra, but sometimes also neighboring and more remote elements, e.g., Bi for Po, Fe and Al for Pb and Th, respectively.

Yet another type of non-isotopic carrier for trace elements is represented by substances which chemically differ from them completely, but have the ability to adsorb them on their surface (Chapt. 4.4.3). Separation of macro- and microcomponent is in such cases most convenient method. The carrier for radioactive gases may be nitrogen or other gases. Organic solvents may serve as carrier in solvent extractions.

In work with radioactive elements, carriers serve several purposes. They often fulfill the function of the so-called *direct carrier*, i.e., a carrier solely for the radioactive element which is the object of our interest. If the system contains several chemically different radioactive elements, a suitable carrier must be chosen for each of them. The specificity and selectivity of a carrier depends on the chemical properties of the other radioactive elements simultaneously present, and on the nature of the relevant chemical operation. In many cases a carrier need not be added separately for each of the radioactive elements present, for several elements may have, under the given circumstances, a similar relationship towards the carrier. One particular substance a macroquantity may then serve as carrier to a certain number, or even a whole series of different radioactive elements. These are the so-called *group* (*collector*) *carriers*. This effect is characteristic particularly for precipitates with a large specific surface, such as $Fe(OH)_3$, MnO_2, etc. On the other hand, if entrainment of other foreign radioactive trace elements by the carriers involved is undesirable, then it becomes necessary to add carriers to the solution not only for elements that are to be excluded, but likewise for those that remain. We then speak of the so-called *hold-back* (*indirect*) *carriers*.

The quantity of a carrier applied depends on the scale of the operation. Most frequently 1–20 mg quantities are applied. Separation from large volumes will require essentially higher quantities of the carrier, up to about 500–600 mg. Amounts smaller than 1 mg are used only in special cases. If the system involves more complicated, multi-step operations, a larger amount of carrier substance will generally be required than for simpler processes. An important role in determining the quantity of a carrier may be played by the radiation intensity, which indicates the requirements for the specific activity of the given radioactive sample.

4.5 Physical Methods of Radionuclide Identification

In radioanalytical work we are often faced with the task of identifying a radionuclide or a mixture of several radionuclides present in the radioactive sample. This is significant, for example, in radionuclide purity control, in activation analysis and in many other procedures. Identifications by physical methods, are of particular advantage because they are non-destructive.

A radioactive nuclide is unequivocally determined by three physical characteristics: (1) the type (or types) of the emitted radiation, (2) half-life and (3) the energy of emitted radiation. In practice, the measurement of half-life and the energy of the radiation may serve in simpler cases to identify a large number of nuclides. In more complex mixtures, however, difficulties are encountered and a systematic identification requires a combination of physical methods with chemical separation and identification.

4.5.1 Determination of the Type of Radiation

The type of a radionuclide's radiation may be determined unambiguously through a study of its penetration in the atmosphere and other substances, such as, plastic materials, aluminum, lead, etc. The following example will illustrate the procedure:

a) A plastic material, 5–7 mm in thickness, is placed between the radioactive sample and the gamma crystal of a scintillation detector. Only gamma radiation (including a small proportion of "Bremsstrahlung") can penetrate a plastic layer of the given thickness. Alpha and beta radiations will be absorbed in the plastic.

b) The radioactive sample is placed at a distance of 6–10 cm from the end-window of a GM tube and the counting rate is observed by means of a counter, measurement being made with and without a plastic plate placed between the radiation source and the counting GM tube. The latter has an efficiency of only 1–2 % for gamma radiation, and at a distance of 8–10 cm alpha radiation is completely absorbed in air and in the window of the tube and cannot be detected. A substantial increase in the counting rate when the plate is removed testifies to the presence of beta radiation. It should, nevertheless, be noted that the very soft beta radiation is emitted for instance, by radionuclides ^{3}H, ^{14}C and ^{63}Ni, will not be detected by this method.

c) Alpha radiation is usually detected by means of scintillation equipment with ZnS, or a proportional counter with discrimination of beta and gamma radiations.

The above procedure is, however, but one of several possibilities. The choice of a method to determine the type of radiation will often depend on the availability of the measuring equipment. Proof of the presence of gamma radiation through the absorption method becomes more complicated when annihilation radiation or X-ray radiation due to *K*-capture may take place along with an emission of gamma radiation.

4.5.2 Determination of Half-Life

The method for determining the half-life may be applied to the identification of radionuclides with an arbitrary character of radiation. From the half-life obtained, we may also determine the value of the decay constant (see Eq. 2.4).

The half-life may be determined from the decay curve which represents the graphical correlation between activity and time. It is best to plot it in semilogarithmic coordinates. In the case of a pure radionuclide sample this correlation is a straight line and the half-life can be easily obtained from it (see Fig. 4.1). In order to reduce errors to the minimum in half-life determination it is advisable to determine the regression curve from experimental points by the least-squares fit.

If the logarithmic correlation between activity and time is a curve instead of a straight line, this testifies to the presence of more than one radionuclide in the sample. If, for instance, two independently decaying radionuclides with different half-lives are present, the decrease in the counting rate (activity in time) may be expressed by the relation

$$I = I_1 + I_2 = I_{01}\,e^{-\lambda_1 t} + I_{02}\,e^{-\lambda_2 t}, \tag{4.14}$$

whose graphical illustration is expressed in curve 1 in Fig. 4.2. After a certain time, the short-lived radionuclide will almost totally disintegrate. From that time the logarithmic correlation takes on a linear course which corresponds solely to the long-lived radionuclide. An extrapolation to the co-ordinates, axis ($t = 0$) will there-

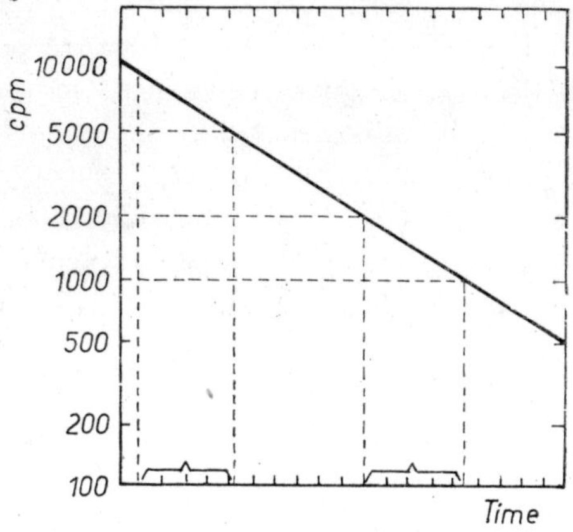

Fig. 4.1. Determination of the half-life from decay curve plotted on semilogarithmic coordinates.

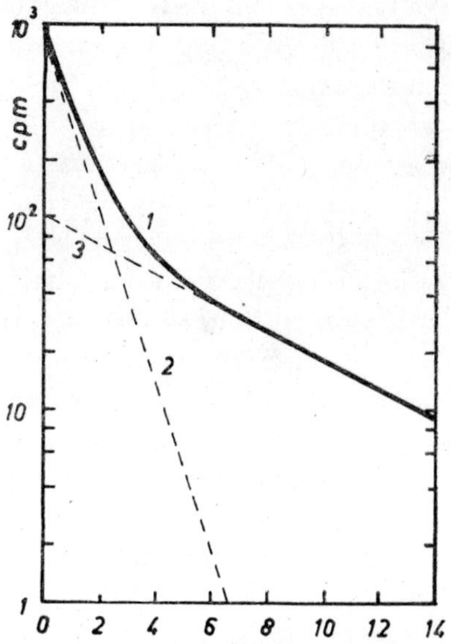

Fig. 4.2. Graphical analysis of a compound decay curve (1 — compound decay curve, 2 — decay curve of a short-lived component, 3 — decay curve of a long-lived component).

fore yield the decay curve for this second radionuclide, which is governed by the relation $I_2 = I_{02}\,e^{-\lambda_2 t}$. By means of a graphical analysis, i.e., by subtraction of this partial decay curve from the total curve, we obtain the decay curve corresponding to the short-lived radionuclide ($I_1 = I - I_{02}\,e^{-\lambda_2 t}$).

If the above analysis of the original decay curve, instead of a straight line as in Fig. 4.2 again yields a curve, the latter is further subjected to a graphical analysis. If the number of radionuclides in the sample is higher than three, this graphical analysis becomes inaccurate. Such complex cases may often be simplified by placing absorbers of various thicknesses between the radiation source and the detector. A basic condition of the applicability of this method even with two radionuclides is that there be a sufficiently great difference in their half-lives (greater than 2 : 1). Moreover, the applicability of this method is, in practice, limited to radionuclides with half-lives ranging from a few tenths of a second to a few weeks, sometimes a few months, which are related with the time factor necessary for obtaining the final results. In the case of short-lived radionuclides (with a half-life shorter than 10 minutes), it will be found more convenient to employ recorders that will assure a continuous recording of the activity. But even these results are markedly affected by statistical fluctuation. A more suitable recording will be that made with a tape-recorder.

Data obtained from such short-lived radionuclides are more conveniently processed in the form of an "integral" decay curve [49, 50]. The radiation intensity is recorded until it falls to the background value. After relevant corrections, the differences are ascertained between the total number of decays I and the number of decays I_t related to the given time t, which are proportional to the amount of the undecayed nuclei. This difference $I-I_t$ is plotted as a function of time on a semilogarithmic scale.

The half-life of short-lived radionuclides is determined with special equipment to transport the sample as quickly as possible from the place of radioactivation to the radiation detector. The most common equipment in use is a semi-automated device working on the pneumatic principle.

When the half-lives are longer, they are determined as a rule by the measurement of their absolute specific activity. This permits the attainment of the proportion of atoms decaying per time unit. From this we obtain the value of the decay constant, and thus we can calculate the half-life. To this end we may also utilize the *Geiger-Nuttal relation* (2.34).

The mutual relationship for beta sources between maximum energy and decay constant was approximated by Sargent [60] in the relation

$$\lambda = kE_{\max}^5, \tag{4.15}$$

where k is a constant dependent upon the character of beta decay. This method is very inaccurate for the determination of half-lives of beta emitters.

4.5.3 Energy Determination of Individual Types of Radioactive Radiation

The energy of a given type of radiation belongs among the most important data in the identification of radionuclides. We shall now review the methods of energy determination for various types of radiation.

4.5.3.1 Alpha Radiation Energy

The simplest way to determine the energy of alpha particles is on the basis of their absorption, with their range being simultaneously determined in either the atmosphere, or some other gas, or in thin aluminum foil. When measuring the intensity of collimated monoenergetic alpha radiation as a function of the distance from the source of radiation, we obtain the relation illustrated in Fig. 4.3, curve A. Despite the great number of collisions involved, including energy loss, the number of alpha particles remains practically constant along the path, owing to their high mass and the ensuing straight line motion. Nevertheless, all the alpha particles do not possess an exactly equal range which would necessarily become manifest in a sudden decline of the number of particles. In reality, the alpha particles with the same initial energy have somewhat differing ranges, which is apparent in a gradual decline of their

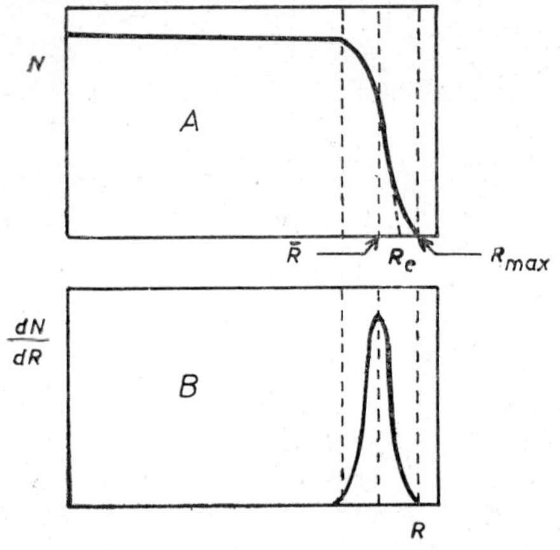

Fig. 4.3. Number of alpha particles of a point source as a function of distance (A) and representation of the range distribution by means of this function (B). ($\bar{R}$ — mean range, R_e — extrapolated range, R_{max} — maximum range.)

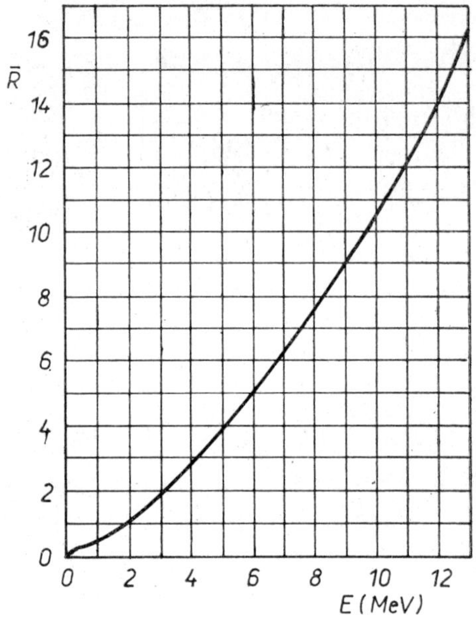

Fig. 4.4. Mean alpha range in the atmosphere $\bar{R}$ as a function of energy.

number. A graphical derivation of curve A yields curve B which represents the distribution of alpha particles according to their range. The peak of this curve corresponding to the inflection point on curve A determines the mean range $\bar{R}$. The fact that mono-energetic alpha particles have different ranges may be explained by random fluctuations in the number of interacting events per unit length of path, and energy losses during interaction. An extrapolation of the tangent passing through the inflection point A to the axis of the abscissa gives the value of the so-called extrapolated range R_e. The final intersection of curve A with the axis of the abscissa is taken as the maximum range R_{max}. The value that can be determined with the greatest precision is that of the central point; consequently it is usual to state this point in determining the alpha radiation energy. Its value may then be determined by means of known relations or graphical correlations. The mean range of alpha particles in air at 760 mm Hg and 15 °C is illustrated in Fig. 4.4 [51].

The range of alpha particles is often expressed also by absorber thickness (mg cm^{-2}) [52]. The problem of alpha range and of the range of other heavily charged particles is treated of in more detail in Ref. [53 and 54].

When greater precision in the energy determination is required, scintillation counters may be used (e.g., scintillation counter with a liquid scintillator), which are calibrated by means of alpha sources. The energy of the unknown alpha particles is then determined by means of the standard method of pulse analysis. The use of spectrometers with a high resolving power enables the energy of individual components to be determined even in mixed alpha sources.

In addition to the above two methods, we may also use proportional and semiconductor counters for an exact determination of alpha particle energy. Furthermore, it may be determined on the basis of alpha particle deviation in the magnetic field, by means of pulse ionization chambers and other methods.

Spectrometry in general is dealt with in Chapt. 3 and the principles of alpha spectrometry are summarized in Ref. [55].

4.5.3.2 Beta Radiation Energy

Energy of beta radiation may be determined with a high degree of precision by means of magnetic spectrometers [55]. Consequently, this method would also be very suitable for the identification of radionuclides. The required equipment is, unfortunately, rather complicated and inaccessible to most laboratories. Therefore, energy of beta particles is determined by other methods based on their absorption. Since beta radiation emitted by radionuclides has a polyenergetic spectrum, maximum energy of this spectrum may serve as the characteristic value. Sometimes the mean energy may be used too.

In the case of beta radiation, it is possible to define exactly the depths of penetration into a particular material at a given radiation energy. The mutual correlation between range and maximum energy is described by a number of empiric formula, and those most frequently employed are summarized in Table 4.3. Many of the above relations overlap in certain areas of energy and yield rather marked deviations in maximum energy values for the same range. Hence, it is best to express this correlation graphically.

Aluminum is a common absorption material in beta absorption studies. The beta range expressed in absorber thickness does not show a close correlation with the atomic number. In the case of copper it is somewhat higher than in aluminum [67].

Table 4.3

Empirical Relationships between the Range and the Maximum Energy of Beta Particles

Area of application		$R_{max} = f(E_{max})$	$E_{max} = f(R_{max})$	References
E_{max} (MeV)	R_{max} (g cm^{-2})			
$0.05 < E_{max} < 0.2$	$0.007 < R_{max} < 0.0012$	$R_{max} = 0.67E_{max}^{5/3}$	$E_{max} = 1.27R_{max}^{0.6}$	[56, 57]
$0.15 < E_{max} < 0.8$	$0.03 < R_{max} < 0.3$	$R_{max} = 0.407E_{max}^{1.38}$	$E_{max} = 1.92R_{max}^{0.725}$	[58]
$E_{max} > 0.8$	$R_{max} > 0.3$	$R_{max} = 0.542E_{max} - 0.133$	$E_{max} = 1.85R_{max} + 0.245$	[58]
$E_{max} > 0.7$	$R_{max} > 0.2$	$R_{max} = 0.543E_{max} - 160$	$E_{max} = 1.84R_{max} + 0.295$	[59]
$E_{max} > 0.6$	$R_{max} > 0.2$	$R_{max} = 0.526E_{max} - 0.094$	$E_{max} = 1.90R_{max} + 0.179$	[60]
$0.05 < E_{max} < 3$	$0.003 < R_{max} < 1.5$	$R_{max} = -0.11 + \sqrt{0.0121E_{max} + 0.27E_{max}^2}$	$E_{max}^2 = 3.7R_{max}^2 + 0.22R_{max}$	[62]
$E_{max} > 1.2$	$R_{max} > 0.4$	$R_{max} = 0.571E_{max} - 0.161$	$E_{max} = 1.75R_{max} + 0.281$	[63]
$0.6 < E_{max} < 1.2$	$0.32 < R_{max} < 0.65$	$R_{max} = 0.546E_{max} - 0.035$	$E_{max} = 1.83R_{max} + 0.064$	[64]

The attenuation of a beam of beta particles in the absorber may be described approximately by the exponential law (relation 2.38). Also, the relationship between the absorption coefficient and maximum energy is of interest, it may likewise be expressed by an empirical formula. The extent of their validity is affected by the character of the beta spectrum and the atomic number of the absorber. In the case of aluminum, we may use *Evans' relation* [65]

$$\mu = \frac{17}{E_{max}^{1.14}} \qquad [\mathrm{cm^2\, g^{-1}}], \tag{4.16}$$

where μ is the mass absorption coefficient, and E_{max} the maximum energy of beta radiation in MeV. For other absorbers we may use *Tittle's equation* [66]

$$\mu = \frac{35\,Z}{AE_{max}^{1.14}} \quad (Z \leqq 13); \qquad \mu = \frac{7.7\,Z^{0.31}}{E_{max}^{1.14}} \quad (Z \geqq 13), \tag{4.17}$$

where A is the atomic weight of the absorber.

The intensity of the detected radiation is considerably affected by the position of the absorber. It increases as the absorber is displaced from the detector towards the source of radiation. Similarly, the effect of detection geometry must likewise be taken into acount.

A Simple Method for Total and Partial Absorption

Beta emitters are often identified by measuring their maximum range in aluminum absorber. Absorbers in increasing thickness are interposed between the detector and the radiation source, until almost all the beta particles are absorbed. With decreasing counts per unit of time, particularly in the last phase when the measured values approach that of the background, the time of measurement must be considerably prolonged to obtain fairly accurate results. The results are figured on a diagram by plotting the log-intensity (cpm) of transmitted radiation as a function of the absorber thickness (mg cm^{-2}). The absorption of the radiation in the counter window and in air should be taken into account. If the absorption curve plotted semilogarithmically is approximately linear or shows a moderate but uniform declination throughout, then only one type of beta radiation is emitted by the sample. The maximum range is determined from the intersection of the absorption curve with a straight line drawn parallel to the abscissa at the level of the background. Maximum energy value may be derived from the value found for the maximum range from the relations in Table 4.3, or the diagram of correlation between range and maximum energy, shown in Fig. 4.5.

The existence of composite beta radiation (complex decay or presence of several radionuclides in the sample) manifests itself on the absorption curve, plotted semilogarithmically by a visible curvature (Fig. 4.6). The absorption curve is subjected to a graphic analysis in a manner similar to the decay curve in the presence of several independently disintegrating radionuclides. After its decomposition into individual components, the range is determined for each of these individually.

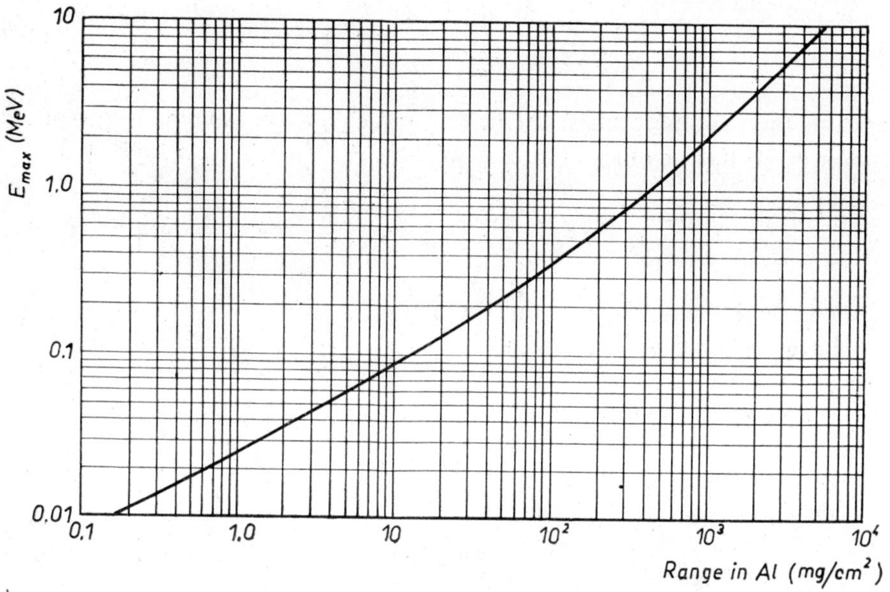

Fig. 4.5. Range of beta particles in aluminum.

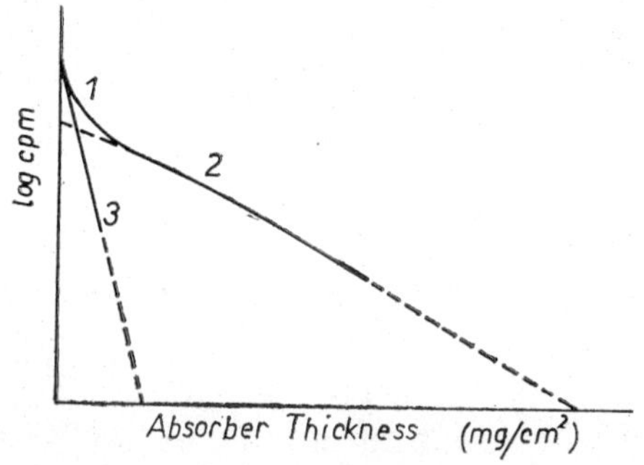

Fig. 4.6. Absorption curve of compound beta radiation and its graphical analysis (1 — compound absorption curve, 2 — absorption curve of a component with higher energy, 3 — absorption curve of a component with lower energy).

The method of partial absorption requires determination only of the initial part of the curve, until the original intensity drops approximately 75%. From the absorption curve we read off the half-thickness value $R_{\frac{1}{2}}$, and from the correlation diagrams $E_{max} = f(R_{\frac{1}{2}})$ we determine the maximum energy directly. From the half-thickness we may first determine the absorption coefficient according to relations (2.39) and (2.40), or generally from a definite sector of the absorption curve, according to the relation

$$\mu = \frac{\ln I_1/I_2}{R_2 - R_1}, \tag{4.18}$$

where I_1 and I_2 represent the counting rates corresponding to two different thicknesses of the absorbers R_1 and R_2, respectively. Maximum energy of the investigated radiation may then be simply determined with the help of relations (4.16) and (4.17), or the graphical correlation $E_{max} = f(\mu)$.

The measured relation between energy and absorption coefficient is relatively inaccurate and this, of course, renders the present method inaccurate, too. The determination of the half-thickness is considerably affected by the character of the spectrum and experimental conditions. In view of these factors, the present method cannot be recommended for maximum energy determination in cases where greater precision is required.

The results in both these simple absorption methods may be influenced by several factors. Very often even the determination of the maximum range is not a simple task. There is a large number of beta radionuclides that simultaneously emit also gamma radiation. This, too, is recorded by the detector; nevertheless, its share in the total counts does not remain constant with increasing absorber thickness. When the absorber thickness approaches the range, measurement of the very low intensity of beta radiation over a high background gamma radiation is extremely difficult. On the other hand, gamma radiation distorts the final part of the curve and incorrect values for the maximum range are obtained if the share of gamma radiation is not subtracted. A similar course of events is induced in positron beta emitters by "annihilation" radiation. Measurement is affected in a like manner, though to a lesser degree, by "Bremsstrahlung" which is produced during electron passage through the substance. Another complicating factor may be the presence of several beta components in the emitted radiation, the presence of conversion electrons, and an anomalous course of the beta spectrum. In every case, however simple, the marked decrease in the counting rate near the complete range remains a problem, and may lead to considerable error.

In complicated cases, maximum range of beta particles may be obtained from experimental absorption curves by several methods.

Feather Comparative Method

The Feather analysis [59, 61, 67] permits the range of beta particles of any arbitrary radionuclide to be expressed in relation to a standard source even in the presence of a gamma or another electromagnetic radiation, and even for soft beta sources in which the preceding simple procedure may have failed. Feather's method is based on the assumption that the intensity of beta radiation of two compared radionuclides decreases by the same ratio when the absorber thickness carries an equal ratio of the range for the two radiations. Absorption measurements are carried out under identical conditions for both specimen and investigated source of radiation. Experimental results are corrected for background and, in case gamma radiation is also present, gamma activity is subtracted from the total activity. Absorption curves are constructed on semilogarithmic coordinates, account being taken of absorption

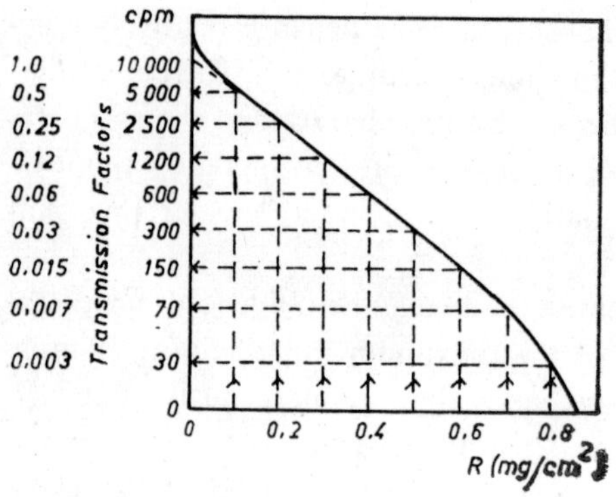

Fig. 4.7. Feather analysis. Calculation of transmission factors from a known absorption curve.

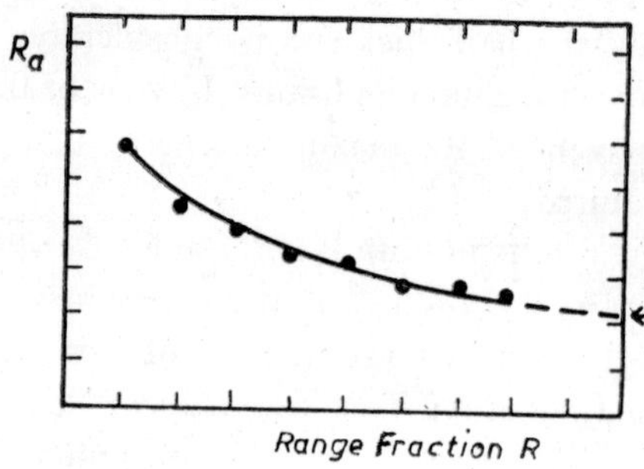

Fig. 4.8. Correlation between apparent range R_a and range quotient, and extrapolation to true range R_t.

in the layer of air and in the counter window. Both the curves are extrapolated to the zero absorber thickness. The tabled value of maximum range for beta radiation of the standard radionuclide is plotted on the abscissa and the absorption curve is extrapolated on the axis of the abscissa to this value. The sector $0R_{max}$ on the abscissa is divided into a certain number of segments, e.g., 10. From each 0.1 R_{max} interval on the abscissa, a straight line is drawn on to the absorption curve and from there, horizontally up to the axis of the ordinates, as shown in Fig. 4.7. From the counting rates corresponding to the intersections on the ordinate axis, the transmission factors are calculated for the absorption thicknesses that correspond to the appropriate portions from the maximum range. The extrapolated counting rate of the unknown radionuclide is multiplied by these transmission factors. The values obtained are plotted on the ordinate axis and horizontal straight lines are drawn from them to the absorption curve. The points of intersection correspond to individual ratios of the range. From the values obtained, the so-called apparent range, R_a, is expressed. The true range, R_t, is then determined by means of Feather's diagram (Fig. 4.8) by extrapolation.

Other graphical interpretations of this comparative method are also possible. Spicyn [69] reports a very simple one. The standard curve is compared with the one under investigation in a graphical illustration on simple millimeter paper

(Fig. 4.9). To this end the standard curve is constructed in the first quadrant and the investigated one in the third. All the values on the axis are calculated as positive. Here, too, the standard curve is extrapolated to the tabled values of the range. The sector on axis 0-A is divided into 10 equal segments. By joining the points correspond-

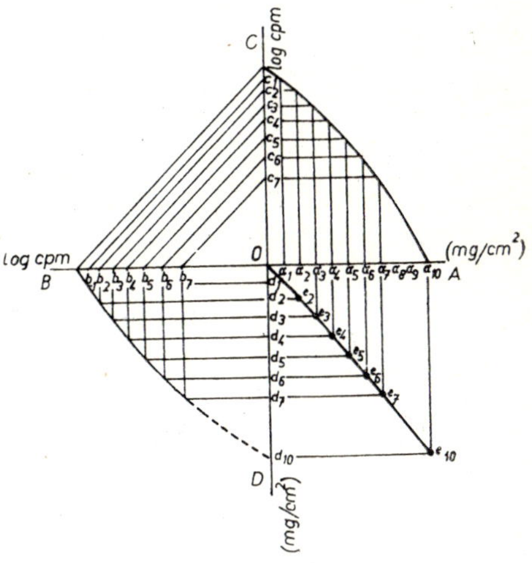

Fig. 4.9. Determination of beta maximum energy by means of the comparative method according to Spicyn et al. [69].

ing to the maximum count rate of the standard and the investigated radiation 0-B and by drawing parallels in the manner indicated, we obtain intersections in the fourth quadrant. The curve is drawn through these points and is extrapolated to point e_{10}, and the required maximum range is found on axis D.

The method according to Feather yields satisfactory results even when the radiation is made up of several components, of course on condition that a certain difference exists between their energies. This method is utilized to determine the maximum energy of individual components following decomposition of the total absorption curve.

A point of interest in the use of this method concerns the comparison of the shape of absorption curves which is known to depend not only on experimental conditions, but also on the character of the beta spectrum. Hence, it is recommended that compared curves be similar in shape and be selected a beta source whose maximum energy does not differ from that of the investigated radiation by more than a factor of two.

Bleuler's and Zünti's Method

A somewhat more complex procedure for determining maximum beta radiation energy is that of Bleuler and Zünti [51, 63, 67], which is based on the construction of theoretical absorption curves for beta spectra with various maximum energies.

This method, too, involves the element of comparison, but instead of dividing into equal segments the sector on the abscissa corresponding to the maximum range, as in Feather's method, here we divide the sector on the ordinate corresponding to the

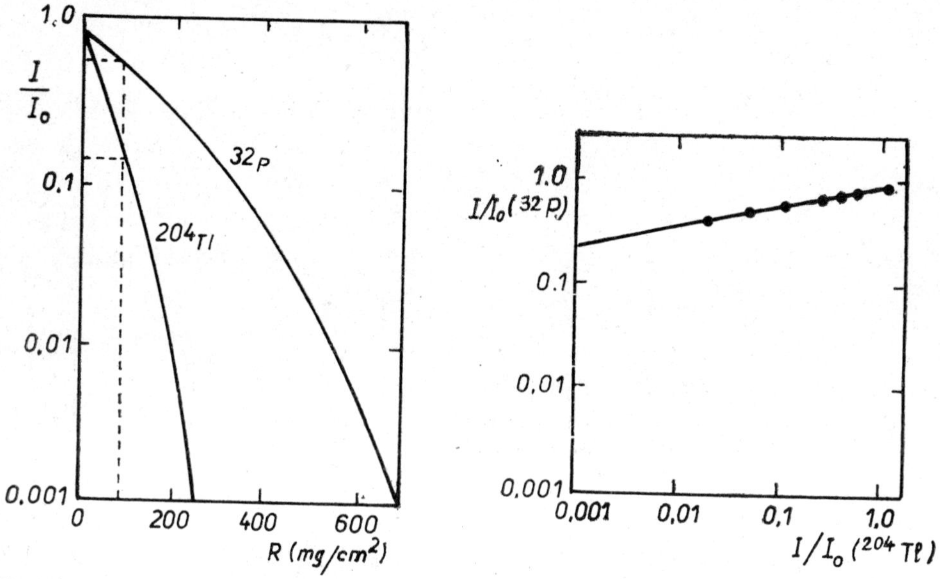

Fig. 4.10A. Comparison of the degree of decline in beta radiation intensity of two radiation sources at one and the same absorber thickness. B. Logarithmus of the transmission factor of the investigated radiation from that of standard radiation.

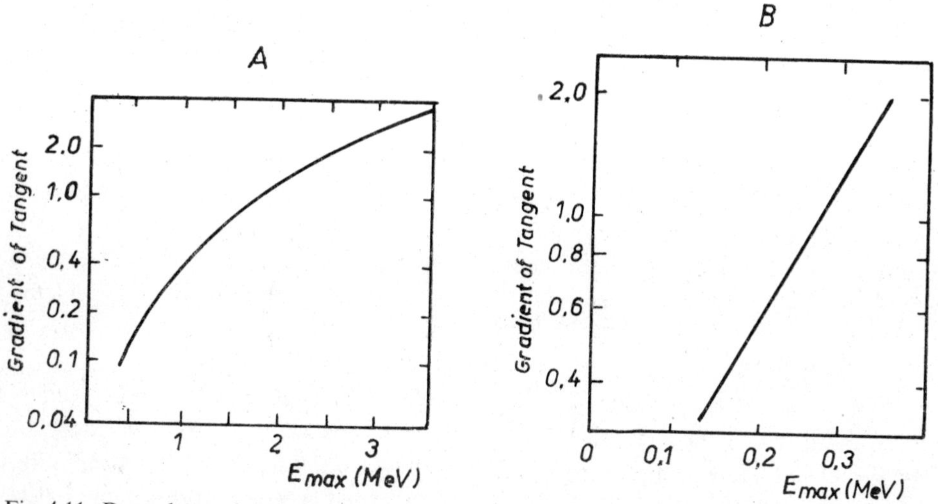

Fig. 4.11. Dependence of slope (gradient of tangent) of logarithmic absorption curves under comparison, and the maximum radiation energy (*A* — slope related to ^{32}P, *B* — slope related to ^{45}Ca).

logarithm of the counting rate, so that every ratio on the abscissa corresponds to the twice-lower radiation intensity passing through the absorber. To these points corresponds the thickness value of the absorber R_n, which may be expressed as

$$R_n = \frac{I_0}{2^n}, \tag{4.19}$$

where I_0 is the radiation intensity without an absorber, and $n = 1, 2, \ldots$ In its essence this method is based on a utilization of the mutual relationship between maximum energy and R_n value, when $n \to \infty$. Often, $E_{max(n)}$ depends so little on n that instead of the limit, we may use the arithmetic mean of all the values of $E_{max(n)}$. However, the R_n value is also affected by the atomic number, which is thus a complicating factor in its application. The curves according to Bleuler and Zünti do not permit the determination of maximum beta spectra energies lower than 0.5 MeV. Nevertheless, this method gives very accurate results.

Harley's and Hallden's Method

In contrast to all the methods cited above for the determination of the maximum energy of beta spectra, that of Harley and Hallden [70, 71] makes use of the initial part of the absorption curve. The analysis is carried through in a simpler manner than in the preceding two methods. Harley's and Hallden's method is based on a comparison of the decline of the relative intensity (I/I_0) of the investigated and the standard radiation, with identical absorber thickness. The graphical correlation is constructed between the logarithm of the transmission factor for standard and that for the investigated radiations (Fig. 4.10). When there is no essential difference in the character of the two radiations being compared, this correlation is linear. A close relationship exists between the tangential gradient of this linear correlation and the maximum energy (Fig. 4.11), by virtue of which E_{max} may be determined. The authors of this method have suggested the use of ^{32}P as standard source for medium and higher energies, and ^{45}Ca for low energies. The advantage of the method is that it yields satisfactory results even at low activities of the unknown emitters, as well as with soft emitters.

Otozai-Hayashi's Method

Recently Otozai and Hayashi [72] proposed a new method for the analysis and evaluation of the results of absorption measurements intended to determine the maximum range, or energy of beta spectra. It is based on the use of a special analyzing paper for analysis prepared on the principle of Bleuler-Zünti's method. The use of this paper makes the procedure of maximum range and energy determination as simple as that with the simplest "absolute" method. An example of the use of this special analyzing paper is shown in Fig. 4.12. The results of absorption measurements for aluminum are immediately plotted on the paper and the maximum ranges and maxi-

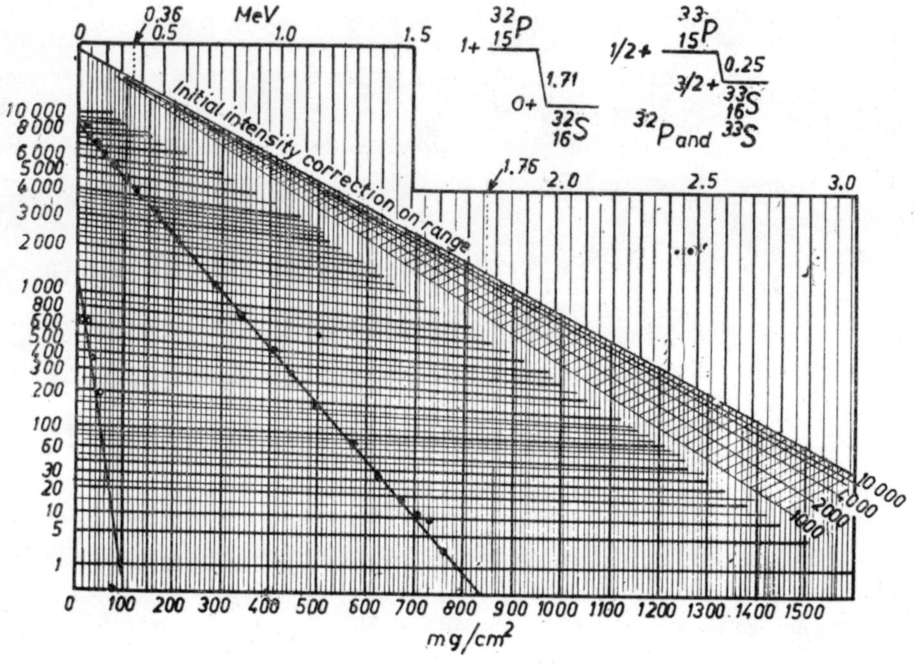

Fig. 4.12. Example of the use of Otozai-Hayashi's special paper in determining maximum energies of the mixed radiation source $^{32}P + {}^{33}P$.

mum energies of the investigated beta radiation are determined by means of linear extrapolation. As seen from the given example, this special paper is also suitable for analysis of radiation consisting of several components. In regard to degree of precision, this method is comparable to those of Feather's or Bleuler's and Zünti's.

A mutual comparison permits these methods to be characterized as follows. The absolute method (method of partial and total absorption) is very simple, but the results may entail considerable error as they are affected by measuring conditions. In the case of the relative methods, the influence of these conditions may be reduced to a minimum, and consequently the results are more accurate, but the procedure is more time-consuming, for it requires that absorption curves also be plotted for standard beta radiation. Methods by which maximum range is not measured, are, as a rule, less accurate (determination E_{max} from absorption coefficients or half-thickness value). An exception to this is the method of Harley and Hallden.

In addition, maximum energy of beta radiation may also be determined by means of scintillation and semiconductor spectrometer.

4.5.3.3 Energy of Gamma Radiation

Like alpha and beta radioactive nuclides, gamma radioactive nuclides may be identified according to the energy of emitted gamma radiation which, if only one radiation component is present, is always characterized by a monoenergetic composition.

Absorption Method

One of the simplest methods for determining the energy of gamma radiation is the absorption method. Absorption of monoenergetic gamma radiation follows the exponential law given by (2.38). This implies that the gamma radiation energy cannot be characterized by the total absorption thickness instead, however, the half absorption thickness $R_{\frac{1}{2}}$, or the absorption coefficient is used (2.44). Absorption thickness is more conveniently expressed in g cm^{-2} and the mass absorption coefficient in cm^2 g^{-1}.

It should be recalled that absorption of gamma radiation follows the exponential law only when certain conditions of measurement are fulfilled. The Compton effect and pair production lead to the formation of secondary electrons which may likewise be detected together with the soft electromagnetic radiation and scattered gamma radiation accompanying these processes. Detection of this secondary radiation may significantly affect the course of absorption curves, thus leading to deviations from the exponential law. Its validity, nevertheless, remains a basic condition in energy determination by the absorption method. Consequently, measurements must be carried out with "good geometry" (Fig. 4.13B), in which gamma radiation from the source is so collimated that only a narrow beam hits the absorber. A colimator is subsequently placed between the detector and the absorber, made of an element with a low atomic number which will prevent scattered quanta and secondary electrons from entering the sensitive volume of the detector. An example of "bad geometry," where secondary electrons are admitted and consequently detected, is shown in Fig. 4.13A. In this case the measured value of the radiation intensity is higher than the correct value.

Absorption materials most commonly used for gamma radiation are lead, copper and aluminum. Lead is used for hard, aluminum for soft gamma radiation. A scintillation or GM counter serves as detector. By placing various absorber thicknesses between the radiation source and the detector and by establishing the correlation

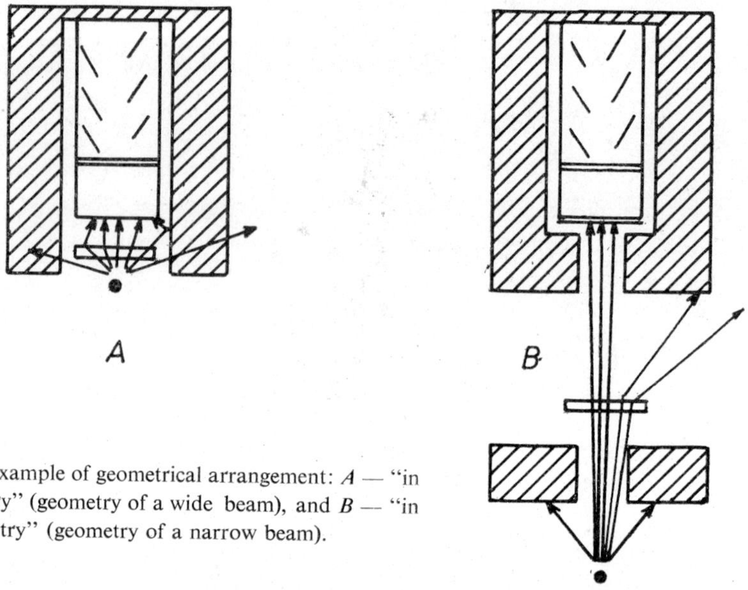

Fig. 4.13. Example of geometrical arrangement: *A* — "in bad geometry" (geometry of a wide beam), and *B* — "in good geometry" (geometry of a narrow beam).

between the logarithm of transmitted radiation intensity and absorber thickness, we obtain absorption curves from which we can obtain the half-thickness value. The latter will mediate the determination of the energy of the investigated gamma radiation through the use of diagrams illustrated in Figs. 4.14 and 4.15. As seen from them, one and the same half-thickness may correspond to two different energy values. The difficulties may be overcome if the absorption curves are plotted for two different absorbers.

Energy of gamma radiation may further be determined from absorption measurements on the basis of correlation diagrams between absorption coefficient and gamma radiation energy (Fig. 4.16). The absorption coefficient may be calculated from the half-thickness according to relation (2.39) or (4.18).

If the gamma radiation is not monoenergetic, the absorption curve may be decomposed and the energy of individual components determined as described for beta radiation (Chapt. 4.5.3.2).

Another important and widely used method for determining gamma radiation energy is scintillation and semiconductor spectrometry. The principles and examples of practical applications are treated in Chapter 3.

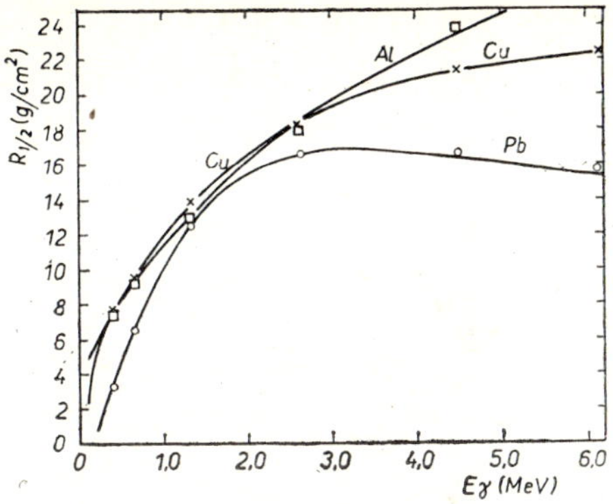

Fig. 4.14. Half absorption thickness for Al, Cu and Pb as an energy function of gamma radiation.

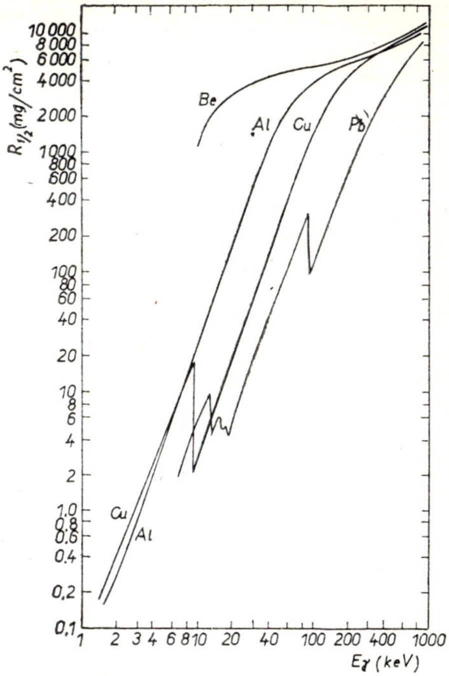

Fig. 4.15. Half absorption thickness for Be, Al, Cu and Pb for low energy gamma radiation.

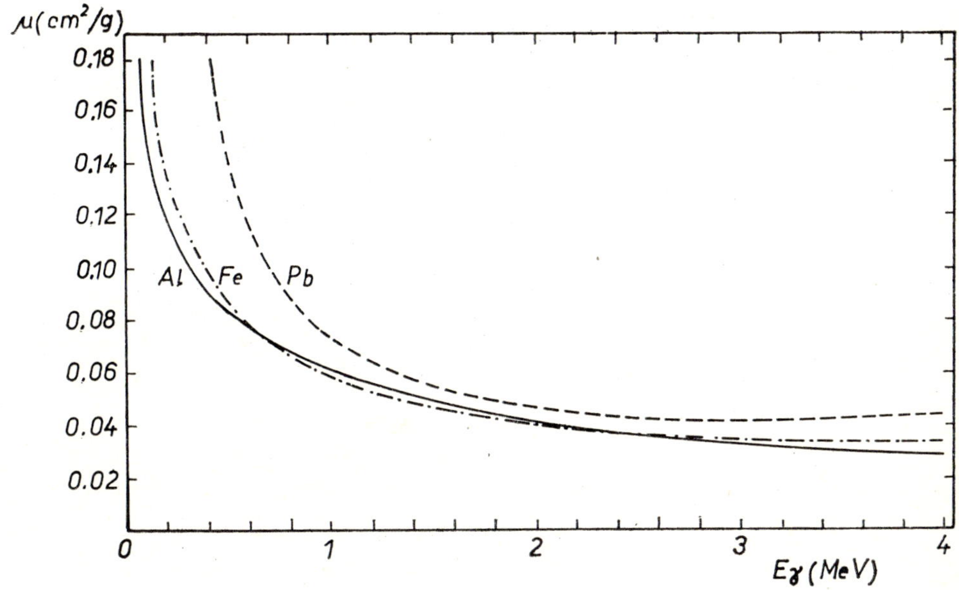

Fig. 4.16. Mass absorption coefficient for Al, Fe and Pb as an energy function of gamma radiation.

Essentially, energy of gamma radiation could be determined by other methods also (e.g., by measuring the energy of secondary electrons, angle of scattering, etc.); these nevertheless cannot compete in simplicity and accuracy with those dealt with above.

4.6 Criteria for the Choice of Suitable Radionuclides

The present treatise deals with all the more important methods of nuclear analytical chemistry. Some of them lay special claims on the choice of the radionuclide. The present chapter will be concerned with criteria for the selection of a suitable radionuclide. These criteria are common for a larger group of radioanalytical methods, but primarily for the selection of a radionuclide which is to serve in tracer methods and in methods based on the interaction of radioactive radiation with the substance being analyzed.

4.6.1 Suitability Criteria for Radioisotopes in the Tracer Method

Radioactive isotopes employed as tracers must meet certain claims which are viewed with reference to the experimental conditions and laboratory facilities. The question of selecting a radioactive or a stable isotope as tracer is, in the case of numerous elements, decided in advance by the number and the properties of existing isotopes. Not every radioactive isotope may be used as a tracer. Certain elements lack suitable radioactive isotopes. Thus, for example, light elements have, for the most part, rather short-lived isotopes, e.g., radioisotopes of He, Li and B have half-lives of less than 2 sec; N, O, Ne, Mg and Al of less than 10 min; F and Si of less than 3 hours (with the exception of ^{32}Si). Therefore, stable isotopes are often used in such cases. If a choice is to be made between a tracer technique with radioactive and one with stable isotopes, for example in hydrogen, carbon or sulphur, preference is generally given to radioactive indication because of its simpler procedure and higher sensitivity. In numerous cases, on the other hand, we may select a radioactive isotope from among several available, e.g., ^{11}C (20 min) and ^{14}C (5760 years), ^{22}Na (2.6 years) and ^{24}Na (15.0 hours), ^{30}P (2.5 min), ^{32}P (14.3 days) and ^{33}P (25 days), ^{126}I (13.3 days), ^{128}I (25 min), ^{130}I (12.6 hours) and ^{131}I (8.0 days), and so on. The selection of a radioactive isotope as tracer takes into account its physical and chemical aspects and also the economic factor.

The Half-Life

The applicability of radioactive isotopes is limited by their lifetime. Isotopes with a very short half-life cannot be utilized in practice as tracers. If the experiment requires a longer time that ten times the half-life of the radioactive isotope, the activity of the latter falls practically below its measurable threshold. Radioactive isotopes with a short half-life may be used only in simple experiments, when rapid methods are employed and when they can be promptly delivered after production.

But difficulties are also encountered in the use of radioisotopes with long half-lives. Most important, long-lived radioactive isotopes emit, as a rule, a low-energy radiation whose detection presents certain experimental troubles and currently requires a larger amount of the radionuclide. The preparation of long-lived radioisotopes is time-consuming and they present problems in safe handling. In view of their considerable half-life they are mostly highly radiotoxic. An eventual contamination is difficult to keep under control as they emit very soft radiation. As indicators, we may use radioactive isotopes with a half-life of a few hours up to many thousand years. For simple experiments it is best to use radioactive isotopes with a half-life of a few days, while long experiments are better served by those with a half-life of several weeks or months.

Type and Energy of Radiation

In selecting a radioactive isotope we must consider also the type and energy of the radiation it emits. Alpha radioactive isotopes are quite unsuitable as alpha radiation has only negligible penetration ability, and this causes difficulties in its detection. Consequently alpha radioactive isotopes are used as indicators only exceptionally.

Practically all the artificial radioactive isotopes are either beta or gamma emitters. Beta radioactive isotopes show certain priorities in comparison with gamma isotopes. Gamma radiation is detected by ionizing methods with a much lower efficiency and is extremely penetrating, and therefore far greater protective measures are required against external radiation. Gamma radioactive isotopes may be used to advantage when detection is to be made through a thicker layer of a material.

Beta radioactive isotopes are of greater significance and beta radiation may be detected with an efficiency many times higher than that of gamma radiation. It is less penetrating and therefore the safety measures during work with beta sources and in their storage do not present any undue difficulties. More convenient are those beta radioactive isotopes that emit hard radiation. The softer the beta radiation emitted, the greater the value of self-absorption and of absorption in air and in the walls or end-window of the tube. Beta radiation with a maximum energy higher than 0.3–0.5 MeV penetrates through the walls of ordinary counting tubes. When the maximum energy does not exceed 0.1–0.3 MeV, tubes with very thin mica windows should be used. If the maximum beta radiation energy is lower than 0.1 MeV, detection is carried out by introducing the samples into the interior of special counters or ionization chambers. This prevents losses that occur when radiation passes through the window of the counting tube. Measurement is made at 2π geometry, whereby efficiency is substantially increased. The difficulties mentioned with GM counter may be obviated by using a liquid scintillator counter for detection. Thus, the type and energy of the detected radiation substantially affect the cost of the measuring equipment.

Facility of Preparation and Specific Activity

The possibility and manner of production and the yield of a radioactive isotope must also be considered, for these constitute important economic factors. The most important source for radioactive isotopes is the nuclear reactor; radioactive isotopes are produced by irradiating the target material with slow neutrons by the (n, γ). In some cases fast neutrons are used and the radioactive isotope is produced by the reaction (n, p) or (n, α). If a particular radioactive isotope cannot be obtained in a reactor, it is prepared by means of accelerators of charged particles. Radioactive isotopes prepared in a nuclear reactor are more accessible and less expensive than those prepared in accelerators. A drawback in the production involving reaction (n, γ) is that the resulting radioactive isotope cannot be separated from the matrix and that its specific activity cannot be arbitrarily increased. Radioactive isotopes with a high specific activity may be obtained in nuclear reactions in which the charge of the nucleus is changed and thereby also the chemical nature of the element. Nuclear reactions of this type are, e.g., (n, p), (d, n), (d, α), etc.

Purity

A radioactive substance used as a tracer must be of radionuclide and radiochemical purity. In view of the importance of purity in nuclear analytical chemistry, a special Chapter 4.3 is devoted to it and to related problems.

Possibility of Labeling Several Elements

When multi-atom molecules are involved, a suitable tracer may often be chosen among radioisotopes of several elements. It is necessary to consider carefully the changes to which the molecule is subject during the proposed operation and the possibility of an exchange reaction taking place.

Such possibilities often arise in the case of labeled compounds. The majority of cases offer the possibility of a choice between ^{14}C or ^{3}H. If tritium does not exchange with protium, and is not split off the molecule during the operation, and if a suitable measuring equipment is available, then it will be preferred, primarily because of its lower cost. If the compound to be labeled contains phosphorus, iodine, bromine, chlorine or sulphur, it is possible and sometimes even suitable to use these elements as tracers.

Radioactive Daughter Products

Certain problems arise in connection with the use of radioisotopes whose disintegration products are also radioactive. For the most part this gives rise to radionuclides with an altered atomic number which consequently show different chemical properties. We endeavor to exclude the interfering effect of daughter products by an appropriate choice of detecting conditions. A chemical separation of the daughter products prior to the use of an indicator would be of value only when the duration of the experiment would be substantially shorter than the half-life of either product. Separation after the test is time-consuming for it must be performed on every sample individually. Daughter products are a complicating factor to a certain extent even when the mother and daughter radionuclides are isotopes of the same element, as is the case, for instance, in nuclear isomers. Hence it is advisable to use, if possible, a radioisotope which disintegrates directly to an inactive nuclide.

In some rare cases the formation of the daughter radioactive nuclide may be utilized also. This has some meaning when the radiation of the original radionuclide, for example, of its low energy, is difficult to detect by current measuring equipment, while on the contrary, radiation emitted by the daughter radionuclide has, from this aspect, superior properties. It is particularly advantageous when the half-life of the daughter product is extremely short and a radioactive equilibrium is achieved in a relatively short time.

Interfering Effects

Certain disturbing effects may arise when radioisotopes are used as tracers. They are:

A. *Isotopic effect.* In heavy elements, the difference in reaction rate and in equilibrium constants due to a difference in isotope masses of the same element, is negligible; however, in the case of light elements, these changes are noticeable. The problem of the isotopic effect is dealt with in more detail in Chapter 5.

B. *Chemical effects of decay and radiation.* When radionuclides are used as tracers, the investigated system is exposed to the effects of radioactive radiation which may evoke in it various chemical changes. It is not possible to present a general and complete picture of the importance of this interfering factor, for yields of radiation chemical reactions depend heavily on the chemical nature of the system. The number of destroyed molecules for each 100 eV energy absorbed is called the *G*-value for the reaction. It is between 1 and 15. Chemical changes due to radiation in the case of these substances are usually so small that they may be ignored. In radiation-sensitive systems, the *G*-value may be as high as 10^6. This is the case when a chain reaction proceeds. Besides the radiation-sensitivity of the substance, radiation-chemical changes are further affected by the value of absorbed energy, which will depend on the type and energy of radiation, and also the value of the specific activity and the time during which the system is exposed to the effects of radiation. For substances highly radiation-sensitive, alpha sources are least suitable, for alpha particles become completely absorbed in the substance and possess a high ionizing ability. Beta sources have a lower ionizing ability and higher penetration, which means that the absorbed energy is lower, and on this account they are more suitable than alpha sources. In view of the fact that gamma radiation is characterized by great penetration and that labeled samples or indicated systems represent in general, minute quantities of substances, the absorbed energy during gamma radiation is only a small fraction of the total radiation energy. Therefore, for radiation-sensitive substances the most suitable sources are those that emit predominantly electromagnetic radiation.

For highly radiation-sensitive substances, we avoid working with very high specific activities and endeavor to process the investigated system in the shortest possible time. By taking into account the above factors in preparing the experiment, it becomes possible in practically every case to reduce the radiation-chemical effects to a negligible minimum. If, however, doubts might arise as to the measure of the disturbing radiation effects, it is advisable to carry out a test trial by irradiating the investigated system, for example, with gamma radiation, and determining the chemical changes by a comparison with a non-irradiated sample.

Even though they are more rare, cases may nevertheless occur when we must take into account also chemical effects that may be induced by radioactive decay and which are designated as internal effects. Every radioactive decay of an element built into a molecule results in chemical changes. These cause a disturbance only when the original molecule contains at least two radioactive atoms. The influence of this effect may then be approximately calculated [22].

The choice of a radioactive isotope influences the degree of the safety measures. From the viewpoint of safe handling we give preference, as far as possible, to less dangerous radioactive isotopes with a shorter half-life, and work with the lowest possible activities. The level of activity and sometimes also the choice of the radioactive isotope depend, to a certain extent, on the sensitivity of the instruments used for detection and recording of radioactive radiation.

4.6.2 Choice of Suitable Radionuclides in Methods Based on the Interaction of Radiation with Matter

The only important aspects to be considered in the choice of the radionuclide serving as the source of radiation, in analytical methods based on the interaction of radiation, are physical and economic. The above methods may be divided essentially, according to the basic principles, into absorption methods, backscattering methods, and methods based on excitation of a new type of radiation.

For obvious reasons, a radionuclide with extremely short half-lives cannot be used as a radiation source. The most suitable radionuclides are those with long half-lives, since with their use the sensitivity of the measuring equipment remains constant over a long period. But difficulties are encountered also when radionuclides with a relatively long half-life are used, for these emit soft radiation which in many cases does not fulfill the energy claims of the method. In view of this, such radionuclides are employed most frequently whose half-lives range between 0.5 and 30 years. Radiation sources are usually utilized until their activity falls below half the value of the original activity, at the most, however, for a period equal to 1.5 T. In some cases, for the sake of maintaining a high degree of accuracy in the measurements, radiation sources are replaced earlier.

The choice of the type of radiation and its hardness depends upon the method utilized and the sample to be analyzed. Because of their short range, alpha emitters are used only in special cases. They may be employed with advantage in gas analyzers that are based on the utilization of ionizing radiation effects and in analysis based on scattering of alpha particles.

The methods most widely used in practice are those that work with beta and gamma sources. The tendency in the beta absorption method is to use radiation sources with the highest possible energy so that the measurement could be carried out at higher thickness values of the analyzed sample. Optimum thickness in the case of soft sources is so small that serious experimental difficulties arise. Similarly, harder beta sources are recommended for the beta backscattering method. With gamma absorption and gamma backscattering methods, on the contrary, soft gamma sources are the more suitable ones.

Both gamma and beta sources can be employed for excitation of X-fluorescent radiation during X-fluorescent analysis. Sources of X-fluorescent radiation excited by gamma radiation show substantial advantages over those excited by beta sources as regards efficiency, purity and convenience for changing target material. To this end we employ the softest possible beta and gamma sources. To obtain "Bremsstrahlung" radiation we may utilize various beta sources in combination with different target materials, in accordance with the required energy of "Bremsstrahlung" radiation.

Beta radioactive nuclides emitting no accompanying radiation are used for the preparation of beta sources. Radiation sources will be dealt with in more detail individually in connection with every particular method in which they are involved. Data on the most frequently used sources of beta, gamma, X, and "Bremsstrahlung" radiation may be found in Ref. [22, 73, and 74].

4.7 Radiological Safety and Radioanalytical Operations

Most operations in radioanalytical practice derive from modifications of current analytical or microanalytical procedures. An essential difference lies in the danger of contamination of the environment when radioactive materials are handled. This chapter includes briefly the most important operations with descriptions of any major modification.

4.7.1 Opening of Ampules

Radioactive substances are usually supplied as solutions in sealed ampules or in vials with special stoppers and metal safety caps. These are opened under a fume hood, in a glove box, or even on a carefully prepared part of a table (if there is no question of volatile or powdered radioactive substances), by means of special long openers which are commercially available. In order to prevent eventual spilling of the substance, ampules or vials are always opened within some container, usually in a beaker placed on a tray. If the special openers are not available, the ampule is held with suitable forceps and opened with an ordinary glass-cutting knife, and in the case of specially sealed ampules, with an ordinary can-opener. The contents of specially sealed ampules may also be removed with a syringe. The contents are transferred by means of a remote-control pipette or Pasteur pipette into containers in which the solution will be stored, or eventually diluted.

Non-volatile solid radioactive materials may be supplied in vials with a safety screw-cap, as used for ordinary chemicals. Opening of containers of powdered radioactive materials presents more danger than of those with liquids, and is done exclusively in dry boxes or under fume hoods. Dilution is carried out by the addition of a suitable amount and type of solvent to the solid radioactive substance. The obtained solution is transferred to a suitable vessel by means of a Pasteur pipette.

4.7.2 Pipetting of Radioactive Solutions

Pipetting enters frequently into work with radioactive substances and as this work often involves extremely minute quantities, micropipettes are often used along with ordinary pipettes. When the radioactive substances amount to 0.001–0.5 ml, the pipettes are usually calibrated "to contain" ("T.C.") and not "to supply" the given volume. Therefore, after emptying the pipette, we wash it with a solvent in order to recover its full content.

In no case may we pipette radioactive solutions by mouth, and similarly we must abstain from handling the pipette by finger. There are several ways of controlling pipettes. Normal-size pipettes may be efficiently controlled by marketed regulators, known as "propipettes" (Fig. 4.17A) which enable pipetting to be done by one hand with an accuracy of 0.01 ml. Another way is to connect the pipette with a suction bulb through a two-way cock as shown in Fig. 4.17B. In opening *1* the bulb is connected with the air, part of the air is pressed out from the bulb and the cock is turned off. The pipette is then immersed into the solution and carefully connected to opening *2*, and when the solution rises in the pipette up to the required mark, the cock is turned off. By repeated manipulation of the cock the pipette may be conveniently emptied. This simple device may be easily set up in any laboratory. Similarly, commercial propipettes can be replaced by home-made safety regulators that consist of a rubber balloon connected to the pipette by a rubber tube containing a glass bead.

To transfer small volumes of radioactive solutions (up to about 2 ml) preference is given to Pasteur's pipettes (Fig. 4.18) which are very handy and relatively safe. Essentially, such a pipette consists of a tube tapering at one end into a fine capillary and closed at the other by a rubber bulb. On a microscale, it may even be used in place of a washing bottle. The bulb volume must always be somewhat larger than that of the pipette.

Pipette control by means of rubber bulbs would, however, lead to considerable inaccuracy. Consequently, they are always connected to special regulators which

A

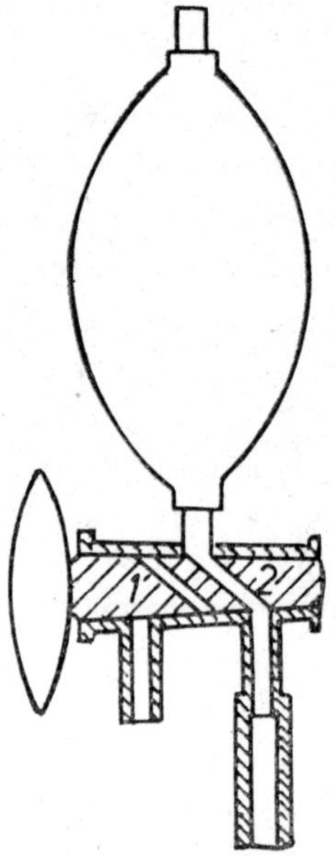

B

Fig. 4.17. *A* — pipette filler (Atomic Accessories, Inc.), *B* — arrangement with a two-way cock for filling and emptying pipettes.

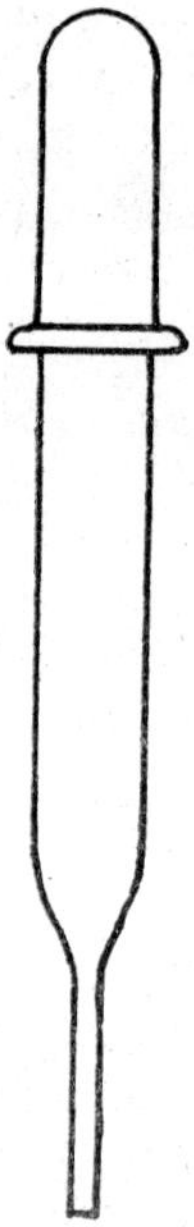

Fig. 4.18. Pasteur's pipette.

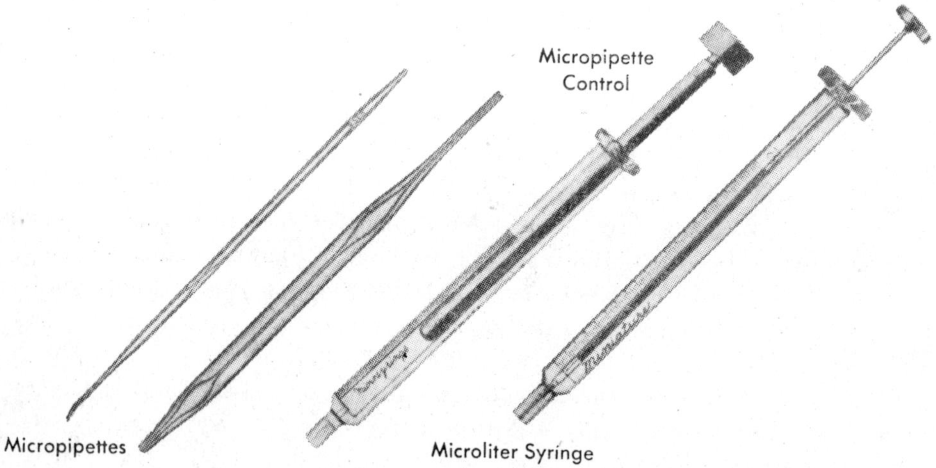

Fig. 4.19. Types of micropipette and micropipette controls (Atomic Accessories, Inc.).

enable pipetting to be done rapidly, exactly and safely. They involve the principle of the syringe, in which the piston can be moved by pulling it or pushing it, but best by turning a plug-head connected with the piston by a screw thread. Some types of regulators with micropipettes are shown in Fig. 4.19.

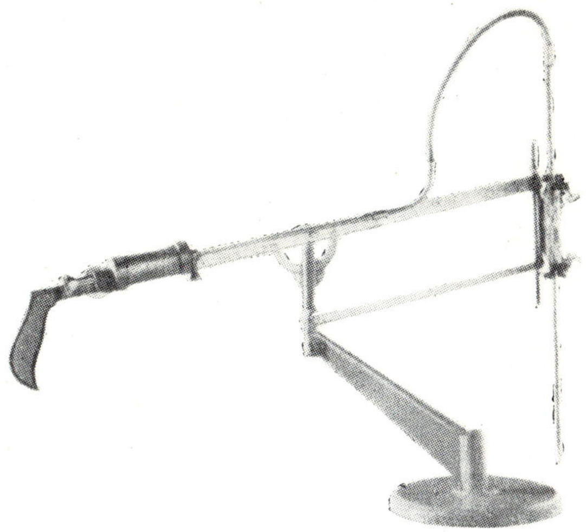

Fig. 4.20. The remote pipetting device.

When handling highly active liquid solutions, we make use of remote control pipette equipment. This permits the manipulator to keep at a safe distance from the unscreened container with the radioactive liquid substance. The pipette or micropipette is placed in the holder with a pantograph-guide system which permits it to be moved in any direction. The pipette is connected by narrow tubing to a syringe fixed at the other end of the holder (Fig. 4.20).

Pipettes (micro- and macropipettes) must be decontaminated as soon as possible after use. They are put away clean and dry. The pipettes are first washed with a suitable decontaminant chosen according to the chemical nature of the solution. The agents most commonly used are hydrochloric acid, chromic acid, nitric acid, citric acid, etc. Finally, the pipette is rinsed with distilled water and dried. A simple device may be employed for micropipette decontamination, consisting of a thin-stemmed funnel connected to the micropipette, which opens into the suction flask. Following decontamination, the pipettes are checked by suitable measuring equipment. If the pipettes cannot be decontaminated immediately after use, they are stored in conical vessels with a layer of a soft material at the bottom so as not to damage the pipette tip. The vessel is then placed on the rack and labeled with the name of the contaminant and eventually with the date when it was used.

4.7.3 Evaporation, Drying and Annealing

These operations do not differ in any great measure from those involving inactive substances. Enhanced safety measures must be taken in order to prevent contamination of the environment by spilling, splashing, or spraying of the handled radioactive materials. Therefore, these operations must be carried out under a good fume hood. As heat source we may use an electric plate, a water bath or a gas burner. In the case of small volumes an infra-red lamp is the most convenient. The degree of heating and rate of evaporation are regulated by altering the distance between the solution and the lamp. Sometimes it will be found convenient to place the container with the solution to be evaporated, under a funnel-shaped vessel whose outlet is connected to a water pump. Care must be taken to prevent the solution from boiling (because of danger of contamination, or formation of aerosols).

Annealing of radioactive materials is carried out in the same way as that of inactive substances.

4.7.4 Precipitation and Crystallization

Among the most widespread and the most important operations involved in radiochemical methods are those based on precipitation and crystallization. Both are often required by the nature of the problem that is being solved with the aid of radioactive tracers; e.g., the preparation of samples for measurement involves the separation of various components, and concentration of radioactive substances involves removal of radiochemical impurities.

If crystallization and precipitation are among those operations to be performed with a substance labeled with a radioactive isotope of the same element as the one being followed, we work for the most part on a macroscale and no unusual difficulties are encountered. The labeled component in the investigated system is, as a rule, present in a macroquantity and the problem is essentially of precipitation and crystallization with the isotope carrier.

The case is more complicated when radioactive elements present in the solution in extremely low concentrations have to be transferred into the solid phase. In such cases precipitation is achieved in two ways: by adding either an isotope carrier or an inert carrier. These types of precipitation obey certain peculiar laws, a knowledge of which is required for successful operation. Precipitation and crystallization conditions—as well as the further processing of the precipitate and solution—will then depend on the manner in which the precipitate is formed, whether the radioelement under investigation passes into the precipitate or remains in the solution.

A major drawback of the precipitation method is that uncontrollable adsorption events and co-precipitation of undesirable admixtures may readily occur. Hence, the working methods must be so selected as to exclude, or at least reduce the co-precipitation. This may be achieved in different ways: e.g., by the use of specific precipitants mostly organic, adding hold-back-carriers, repeating precipitation, adding complex-forming and surface-active substances, decreasing the pH of the solution, etc.

Problems related to precipitation of highly diluted solutions are dealt with at greater length under Chapt. 4.4.3 and 4.4.4.

Some important nuclear analytical methods (e.g., that of isotope dilution) do not require a quantitative recovery of the given element or compound, but lay stress chiefly on the purity of the recovered precipitate of known composition. This extends considerably the choice of the precipitants. Reprecipitation and recrystallization is today a well-known method for achieving higher purity of labeled compounds after their separation. The precipitation and crystallization processes may be repeated until the desired purity is achieved.

4.7.5 Separation of the Solid from the Liquid Phase

The solid phase may be separated from the liquid by filtration or centrifugation.

For filtration we use analytical funnels and filtration cups (with low activities), but more frequently, Büchner funnels are used and particularly a special detachable filtration equipment. The latter is used in particular in the preparation of radioactive samples; a more detailed description of them is given under 4.8.2. This equipment permits a combination of overpressure and exhaustion (suction) whereby rapid filtration is achieved together with uniform distribution of the precipitate over the entire surface of the filter. The pressure difference is chosen according to the nature of the filtred substance and the filter itself. Further advantages of this equipment reside in the ease with which it may be decontaminated and the filter removed. When a Büchner funnel is used we fix the filter paper to the walls of the funnel with a thin paraffin layer. Filtration is accelerated by using a vacuum pump (Fig. 4.21).

If in the separation of the solid from the liquid phase we are primarily concerned with the solution, it will prove more profitable to perform the separation by centrifugation. Its priority over filtration is that no filter paper is employed and the centrifuge vessel is easy to decontaminate. If an ordinary centrifugal vessel is used in this method, and the precipitate is to be further processed, the latter is transferred with the aid of a transferring pipette which does not taper into a capillary. If a quantitative process is necessary, the vessel is washed several times with a small amount of water. Certain types of precipitates are removed with difficulty from the pipette walls. In such cases we dissolve and precipitate it again. Under such circumstances it is far

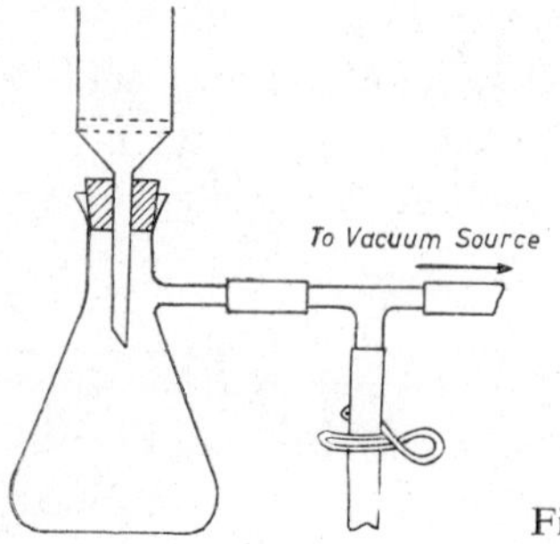

Fig. 4.21. A simple filtrating system.

more convenient to make use of centrifugal vessels with a flat removable bottom. Lately, separation of the solid from the liquid phase has come to be practiced by centrifugal filtration, which embodies a combination of centrifugation and filtration.

Polluted solutions may be filtered directly during their collection by means of special detachable pipettes. During suction the solution passes through a filter paper held between two filters inserts.

Mouth-controlled pipettes should not be used in radioisotope laboratories. Compressible polyethylene syringes, currently employed in chemical laboratories, will often prove entirely satisfactory. If the nature of the solution or its temperature excludes the use of plastic syringes, a glass one may be used with a rubber suction bulb. When work is done on a semimicro- or microscale, smaller syringes with a fine outlet capillary or Pasteur's pipettes are used.

4.7.6 Solvent Extraction

Solvent extraction methods are of extraordinary significance in nuclear analysis for the separation of radionuclides or isolation and concentration of a particular species. These methods are simple and rapid, and can be applied on either a macro- or a microscale (see also 4.3.5). One phase is usually an aqueous solution, the other an organic solvent. The method is employed mostly for the extraction of various metal ions from aqueous solutions into an organic solvent. Most frequently the metal ion is complexed and extracted into an organic solvent. For analytical purposes the so-called chelate extraction systems are used, where a chelate-complex-forming organic compound is dissolved in a solvent of low dielectric constant and the metal ion from the aqueous phase is transferred as a neutral complex into the organic medium [76, 77]. A classification of extracting systems may be found in Ref. [78], and several data of interest and references also in Ref. [50, 79 and 80].

The complexation equilibrium and the equilibrium constant for the process is given by

$$(M^{n+})_a + n(HR)_o \rightleftharpoons (MR_n)_o + n(H^+)_a \tag{4.20}$$

and

$$K_{ext} = \frac{(MR_n)_o(H^+)_a^n}{(M^{n+})_a(HR)_o^n}, \tag{4.21}$$

where indices o and a refer to the organic and the aqueous phase, respectively.

Regarding an uncomplicated extraction process, the distribution is given by

$$D = \frac{(MR_n)_o}{(M^{n+})_a} = K_{ext}\frac{(HR)_o^n}{(H^+)_a^n} = \frac{K_D K_s K_a^n}{K_{ch}^n}\frac{(HR)_o^n}{(H^+)_a^n}, \tag{4.22}$$

where K_D is the distribution coefficient of the chelate between the two phases, K_s the stability constant of chelate, K_a the dissociation constant of the chelate-forming agent, and K_{ch} the distribution coefficient of the chelating agent (HR) between two phases.

Relation (4.22) makes it evident that the distribution ratio is greater (1) the more soluble the chelate in the organic solvent, (2) the higher the stability of the chelate complex, (3) the higher the dissociation constant of the chelating agent, (4) the better the chelating agent soluble in the aqueous phase, (5) the higher the concentration of the chelating agent, and (6) the higher the pH of the aqueous phase. On the basis of these conditions, an extraction system may be selected with the highest distribution ratio possible.

The selection of suitable conditions permits the extraction of a larger number of radioactive or labeled metal ions. On the other hand, the extraction method in many cases yields a satisfactory degree of selectivity, which is achieved by a suitable pH adjustment and the use of a masking agent that prevents the extraction of certain ions into the organic phase.

In practice it is often convenient to express the per cent extraction (E_y) as a measure of the extractability for a given system. The per cent extracted and the distribution ratio are related as follows:

$$E_y = \frac{100D}{D + V_a/V_o}, \tag{4.23}$$

where V_a and V_o are volumes of the aqueous and organic phases, respectively. If the value E_y is known, we may easily calculate whether and how many times the extraction has to be repeated in order to achieve the required separation factor. The extraction is affected also by the kinetic factors. The speed with which the substance is transferred from one phase into the other depends on the diffusion coefficient of the extracted substance and on the intensity of shaking. As a rule, an intensive shaking for a few minutes is sufficient. If the last traces of the substance should be removed from the aqueous phase the extraction should be repeated several times with a small amount of the extracting agent.

If the extraction is to be done in a single process, or is to be repeated only a few times, batch extraction, with an extraction funnel is performed.

The degree of safety measures will depend on the activity level and radiotoxicity of the radionuclide. It should be kept in mind that the chief source of contamination may be leakage around the stopper and cock. For the most part, extraction must be carried out under a fume hood. In any case, separatory funnels are handled with rubber gloves and over a safety vessel. Enhanced care is required in manipulating with the separatory funnel once equilibrium is reached in the case of very volatile solvents, or during extraction at increased temperatures, when overpressure takes place in the funnel which may cause the substance to spray out and contaminate the environment.

When the distribution ratio is low and thus more extractions are needed, and also when the activity of the solution is higher, the extraction is carried out in a continuous extraction apparatus (Fig. 4.22). Extractors in Fig. 4.22*A* and *B* are designed for systems in which the organic solvent is lighter than water [79–81]; *C* and *D* for cases when the solvent is heavier than water [50]. We shall illustrate the course of extraction on the example of the extractor shown in Fig. 4.22*A*. The aqueous solution from which the radioactive substance is to be extracted into the organic solvent forming the upper phase is introduced into the lower part of the extraction apparatus. The beaker is heated and the vapors of the solvent reach the condenser, where they condense into the inner tube. This causes the hydrostatic pressure to increase, as a result of which, part of the solvent permeates through the porous wall into the aqueous solution where it rises to the upper layer. The radioactive substance simultaneously passes from the aqueous solution into the organic solvent. The solution of the radioactive substance in the organic solvent passes through the flow-off tube into the beaker, where it is concentrated. The same procedure is observed in the case of extraction with heavier-than-water solvents.

Work on small volumes is done with tiny extraction vessels made of glass tubing or special microextractors, an example of which is shown in Fig. 4.23.

Sometimes, extraction may be followed directly by activity measurement with the aid of liquid GM counter, scintillation (well-type) counters, etc. However, often an extra step is involved; the solvent is removed so that conditions might be prepared for further separation or activity determination. To this end we either evaporate the organic solvent and oxidize the residual organic complexes or back-extract the ions into the aqueous phase of the acid solution. The recovered extract may sometimes be freed from impurities by back-extraction with water solutions.

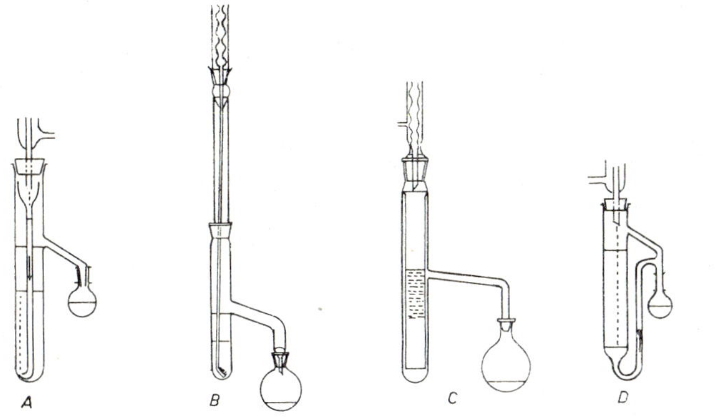

Fig. 4.22. Continuous extractors: *A* and *B* for organic solvents lighter than water, *C* and *D* for heavier than water organic solvents.

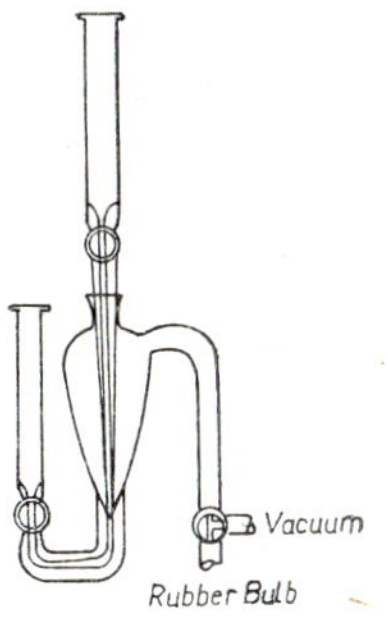

Fig. 4.23. Microextractor.

4.7.7 Electrochemical Methods

Electrochemical separation of radioactive elements is also applied in radiochemical analysis. The electrochemical behavior of an element depends to a considerable extent on the electrolyte composition, the type of instrumentation used (including the material and the shape of electrodes), and further experimental conditions. Electrolysis is one of electrochemical procedure may be used without any external source to remove from the solution the relevant radioactive (labeled) element or even an undesirable admixture. The given element may be removed from the solution with a carrier or without it; in the latter case, however, possible deviations, due to trace concentrations, have to be envisaged (see 4.4.7). The following elements may be plated out from aqueous solutions by cathodic deposition [50]: Ag, Au, Bi, Cd, Co, Cr, Cu, Fe, Ga, Ge, Hg, In, Ir, Mn, Mo, Ni, Pb, Pd, Po, Pt, Re, Rh, Ru, Sb, Se, Sn, Tc, Te and Tl. Elements Ac, Co, Mn, Pb and Po are deposited as oxides on the anode. Halogenides are deposited on a silver anode in the form of the corresponding silver halogenide.

Not all the elements, however, may be removed completely and easily. In macrocale electrolysis an element is considered to be completely removed when its concentration in the solution drops below 10^{-6} M. This also implies the limitation of this method. Thus, it cannot be applied for short-lived radionuclides present in the solution in trace concentrations. The quantity of the element left in the solution after electrolysis may be reduced by deposition under higher voltage, though this is frequently to the detriment of purity of the deposit if the solution also contains other elements. These complications may sometimes be overcome by using complex-forming agents.

Usually platinum is used as a cathode, but mercury is applied when traces are separated by an electrochemical method. The anode may be of platinum, a platinum-iridium (10%) alloy and in special cases, of silver and lead. Since hydrogen has a high overvoltage on a mercury cathode, some heavy radioelements, even from acid aqueous solutions, can be electroseparated; another advantage of mercury is the relatively small change in surface characteristics during separation of the particular element. On the other hand, it is not always a simple matter to recover the isolated element from the mercury. It should, nevertheless, be used every time if the isolated radioelement is an impurity.

Internal electrolytic separation is particularly advantageous in isolating small amounts of an element from a large quantity of another. Great care must be taken in applying this method, to prevent losses through adsorption on the walls of the vessel or through radiocolloid formation.

Relevant literature describes various equipment for electrolytic separation which may be utilized also for radioactive material [80], or special equipment for the electrolysis of radioactive elements [82]. Useful data on this method may also be found under Ref. [83–87].

4.7.8 Other Operations and Methods

The most significant other procedures in radiochemical analysis are chromatographic and ion exchange methods. They will be described in the second volume of this work.

Additionally, distillation, volatilization and other methods are applied occasionally in radiochemical practice.

4.8 Sample Preparation for Radioactivity Measurement

4.8.1 Introduction

Much attention should be given to the preparation of samples for activity measurements. An incorrect or negligent preparation may easily give rise to errors which lead to faulty results and erroneous conclusions even though the whole experiment, had been carried out absolutely correctly.

An indispensable condition for correct preparation of samples is their geometrical and physical homogeneity. The measured value of relative activities will depend on the thickness of the sample layer which determines the value of self-absorption and also of backscattering of the support, and on the size and shape of the surface which has a significant bearing upon the overall geometrical measuring conditions. As correction for surface is a rather complicated operation, samples for relative measurements are always prepared with the same size and shape of surfaces.

Sample preparation should likewise rule out the effect of the layer thickness on the measured values. This is generally required for alpha and beta sources and only occasionally for gamma sources. One way to attain this is to spread the sample in a very thin layer to rule out self-absorption. This procedure is employed mainly when radioactive nuclides emitting hard beta radiation are measured. In the case of alpha and soft beta radiation, extraordinarily thin samples must be prepared. This is rather difficult, in particular as far as a uniform distribution of the sample over the whole area is concerned.

A second way is to provide samples of standard thickness. In this case self-absorption is uniform in all the samples. Nevertheless, sample preparation in this manner is rather exacting.

Samples with very soft beta radiation may also be prepared such that their thickness is larger than the beta range. Here, radiations emitted in the lower layers are absorbed in their passage through the total absorption layer of the sample. The value of the relative activity measured after its passage through the total absorption layer does not depend any more on further increase of thickness, and is proportional solely to the specific activity of the radioactive substance.

If the effect of self-absorption is not excluded by one of these methods during the preparation of the sample, and samples of various thicknesses are measured, a correction must be made by means of the dependence diagram of self-absorption on absorber thickness.

Several techniques have been suggested for sample preparation, and their choice depends on the physical and chemical state of the radioactive material, the type and energy of radiation, the total amount of the radioactive substance, its concentration in the solution, the number of samples that is to be prepared, the required accuracy of measurement, and the type of the applied detector. We shall next deal with the techniques of preparation of samples for counting, with special regard to those requirements which must be taken into account in the preparation of beta radioactive materials.

Radioanalytical methods most frequently involve preparation of samples for measurement of radioactivity from solutions and therefore this case will be described in some detail. Activity measurements may, in principle, be carried out with solid, liquid and gaseous samples.

4.8.2 Solid Samples

Solid samples may be prepared from solutions by evaporation, precipitation or electrolytic deposition, but also from radioactive powdered materials and radioactive gases.

Evaporation of Solutions

Samples may thus be prepared when a non-volatile radioactive substance is dissolved in a volatile solvent. Taking up and spotting of solutions for evaporation is done with micropipettes (described earlier).

Liquid samples are spotted, evaporated and measured on a specially prepared dish or disk, made of glass, aluminum, copper, stainless steel, plastics or other material which is chosen so that the solution will not attack it. Thus, for instance, acids attack aluminum, in some cases even stainless steel and copper. Hence, glass and plastic materials will be suitable for acid solutions. Organic solvents, on the other hand, may attack certain plastics. However, a suitable material may be found for every solution.

In a simple case of sample mounting, a known amount of the solution is transferred by means of a micropipette onto the center of the dish or disk (Fig. 4.24) and the solvent is left to evaporate, for instance, under an infra-red lamp. Some difficulty is encountered in this procedure as regards the uniformity of the deposit and its thickness, due to the viscosity and the concentration of the substance in a ring around the edge. More uniform deposits are obtained when more diluted solutions are evaporated. On the other hand, more diluted solutions must be spotted repeatedly. In this case, the slurry tends to concentrate preferentially at the site where most of it was deposited in the preceding evaporation. This gives rise to aggregates which grow up slowly in the course of evaporation. This simplest mode of sample mounting may be employed without danger of introducing serious errors into the results only when a minute amount of the solution is sufficient and when the deposit forms a very thin

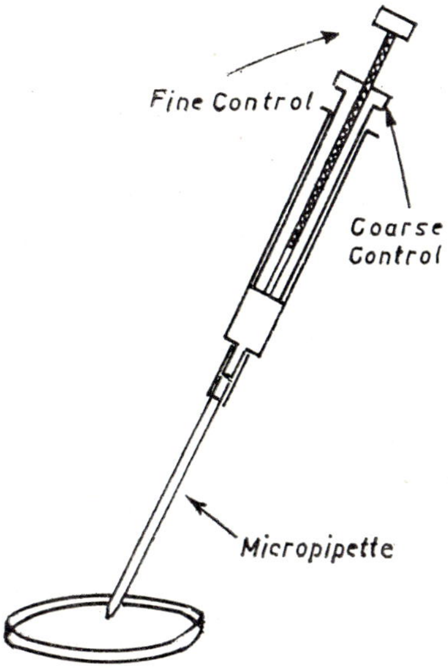

Fig. 4.24. The use of a micropipette with coarse and fine control in sample preparation.

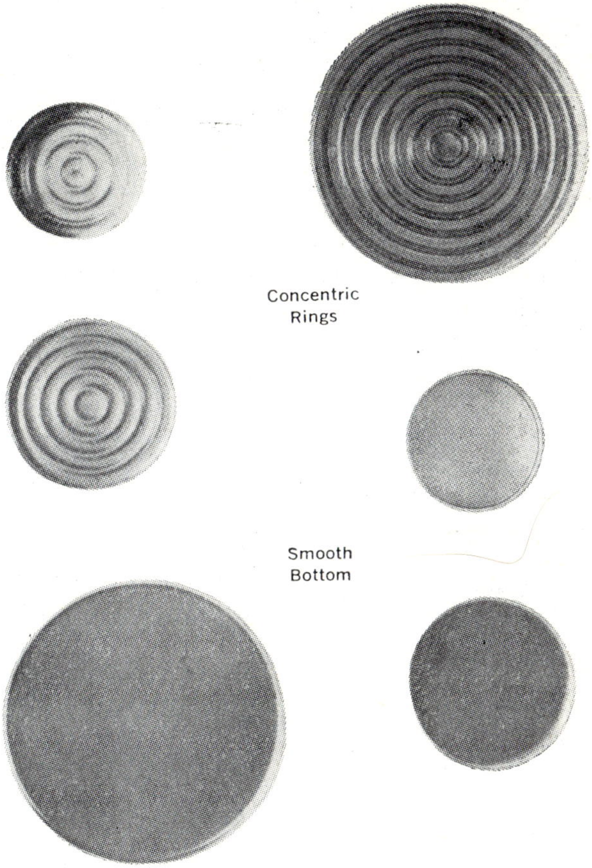

Fig. 4.25. Planchets for sample preparation (Baird Atomic, Inc.).

layer, i.e., during higher specific activities of the solution. These conditions, however, are rarely fulfilled in analytical applications. Hence, various techniques are utilized to rule out or at least substantially reduce the effect of these factors.

When dishes with a smooth bottom are used it will be found helpful to rotate them during drying. The optimum rotating speed differs with every substance. Even better results may be achieved by a combination of rotation with a speeding-up of evaporation by means of a hot air stream, or of vacuum.

A widely used technique to obtain a better distribution of the substance in the deposit, and to improve accuracy and reproducibility in sample mounting, is based on the design of the measuring dish or disk. Various types of these dishes, commercially supplied, are shown in Fig. 4.25.

A series of concentric rings grooved into the bottom of the dish has proved helpful in maintaining a uniform surface of the sample, as well as improving its distribution. If these ready-made dishes are not available, the bottom of ordinary dishes may be smeared with a layer of a suitable wax (of a higher melting point) or a water-resistant cement on which the concentric rings are grooved by means of a pair of compasses.

A uniform surface and thickness of the deposit may often be achieved by evaporating the solution on an absorbing material, which is spread on the bottom of a dish. Filter paper, cigarette paper, glass gauze, etc., may be used for this purposes. The solution is first absorbed by this material and thus spreads over a certain surface—generally the same with equal volumes—and then follows evaporation of solution. Here, however, certain circumstances must be taken into account which are likely to become sources of errors. During evaporation, the solution may permeate from the bottom to the surface of the paper where solid substances concentrate. This considerably lowers self-absorption and also backscattering. The effect of this factor may be reduced by rapid dessication and the use of highly volatile solvents [88]. The measure to which the results may be affected by the diffusion of the solution to the surface of the paper may be ascertained by measuring the sample with the paper turned over to the other side, or by preparing and measuring a series of samples with a constant activity, but with different amounts of the solid inactive substance and solvent.

Well-defined surface and relatively uniform deposits may be achieved by first placing on the dish some other material as deposit in the required shape, e.g., insulin. When it dries, the radioactive solution is transferred onto it: the latter will tend to remain on the previous deposit.

Satisfactory accuracy and reproducibility in the preparation of samples may often be achieved by the use of ordinary dishes on which undesirable spreading of the solution is prevented with the aid, e.g., of Zapon lacquer—with which a definite area is encompassed.

Distribution uniformity of the deposit may be greatly enhanced by the use of wetting agents and other factors that reduce the formation of large crystals. This uniformity of distribution can best be checked by radiography. A special technique has been proposed for mounting samples of substances labeled with ^{14}C or ^{3}H [89, 90]. The sample is dissolved in an aqueous solution of such a substance which, after the evaporation of water, will form a uniformly spread gel mass (carboxymethylcellulose, agar).

When mounting samples by the evaporation method, we must not lose sight of the possibility of losses in the radioactive substance itself during evaporation. Volatility, it should be observed, is a relative term. Losses of so-called non-volatile materials may be considerable when the given substance occurs in a very thin film. Similarly, in some rare cases the radioactive compound may hydrolyze during evaporation forming volatile products.

Sample Mounting by Precipitation

Solid samples may also be prepared for activity measurement from solutions of radioactive materials by precipitating them by the addition of a suitable precipitant and separating the solid from the liquid phase. This method will be of particular advantage when samples are to be prepared from highly diluted solutions that would involve evaporation of large volumes, thus making the process time-consuming and the method less accurate. Addition of the carrier enables the radioactive substance to be precipitated and thereby concentrated into a small volume. Here we may make use also of various types of co-precipitations, mentioned in Chapt. 4.4.3 and 4.4.4. The techniques of precipitation and of separation of the solid from the liquid phase are described in Chapt. 4.7.4 and 4.7.5.

Care must be taken during precipitation to exclude the formation of flocculent, hydrated precipitates which not only crack on drying, but prevent, in particular, a uniform distribution of the precipitate.

Samples from the precipitate are prepared most frequently for measurement by filtration, which generally assures a very uniform distribution of the solid phase. Many filtration equipment has been described in the literature and several types are commercially available. These must fulfill all the requirements of ordinary filters designed for radioactive materials and in addition, must possess a well-defined area on which deposit formation occurs, thus helping its uniform distribution over the entire surface. Use is made nearly exclusively of special demountable equipment made of glass, Teflon (polytetrafluoroethylene resin), or stainless steel. This permits the samples to be rapidly transferred and washed with only a small portion of the washing solution; moreover, the filtration medium with the deposit can be easily removed and the equipment itself readily decontaminated. Some types of such filtration equipment are shown in Fig. 4.26. Filtration is usually carried out under suction, but a combination of suction with overpressure is also possible.

In the simplest cases, filter paper placed over a perforated support serves as the filtration medium. This procedure, however, may give rise to various difficulties in connection with drying of the sample, its transfer from the filtration equipment and preservation of its shape. Its transfer to a solid underlay must be made while the sample is still damp; otherwise it will become deformed. The paper will tend to warp during drying. This may be prevented by holding it down with a weight ring.

Some filter equipment is constructed to allow the filter paper to be placed directly on the bottom-perforated dish. Instead of the filter paper, a glass frit ring or one of plastic material [88, 91] which are more resistant to warping may be used.

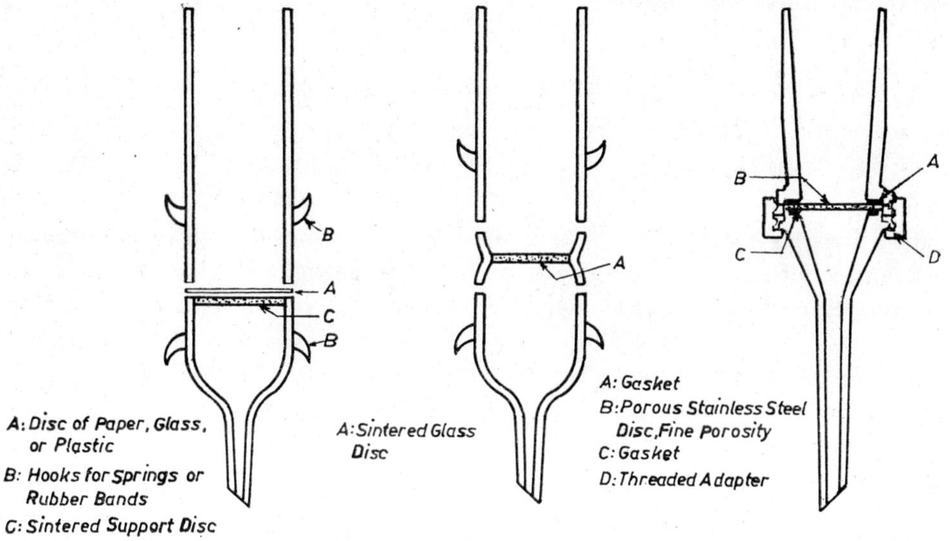

Fig. 4.26. Some types of filtration devices [88].

In view of these drawbacks in flexible filter media, it is generally more convenient to use solid media from which the filtrate need not be removed for measurements may be carried out directly on them. They therefore also fulfill the function of the support (dish) during measurement. These filters are generally made of porous glass or stainless steel.

Before measurement, the prepared samples must be carefully dried (self-absorption and backscattering). However, the filtrate cake tends to crack and deform in this process, thereby rendering the preparation of samples less reproducible, besides leading to possible losses through powdering and even to contamination of the measuring equipment. There are several means for the satisfactory drying of a sample while preventing it from cracking and deforming;

A. A convenient way is to dry it at a lower temperature with the use of an infra-red lamp, or if possible, at room temperature with air-exhaust, or in an desiccator. The last washing should be done with highly volatile liquids (ether, methanol, acetone, etc.) and the sample will then dry within 10–20 min at room temperature without heating. Account should nevertheless be taken of the possibility of the sample absorbing some moisture from the air; this depends primarily on the nature of the sample, and also on the humidity of the air.

B. An important role is played also by the nature of the washing liquid. Some substances swell in contact with certain liquids and this makes the formation of cracks and irregularities more likely. It is advisable to use for the final washing a liquid that dissolves the given substance as little as possible.

C. In the majority of cases several possible precipitating reactions may be selected for the same substance. The precipitates thus formed will then differ according to the size and shape of the crystals. As far as possible we decide that for a precipitating reaction which will yield a satisfactorily large-grained precipitate and permit the selection of suitable washing agents in view of points *A* and *B* above. We should endeavor to avoid reactions that yield flocculent precipitates which adversely affect the mounting of samples for measurement.

D. A satisfactory sample stability may be achieved by the addition of suitable sticking agents in the form of diluted solutions (e.g., Zapon lacquer in alcohol, collodion in acetone). The solutions should be so dissolved that a few drops of them, when added after drying of the sample, will not cause a significant change in the absorber thickness (this depends on radiation energy, a maximum being, however, 0.1 mg cm^{-2}).

E. Sample stability during drying is sometimes also affected by rate of filtration, in the sense that deposits of better quality are obtained under a slow filtration.

When samples are prepared by centrifugation (see 4.7.5), the bottom of the centrifugation cell is formed by the measuring dish, which is demounted following separation of both the phases, and remnants of the solvent are removed by drying. Selection of suitable conditions is governed by the same principles as in the case of filtration.

Electrolytic Deposition

Sometimes it is convenient to prepare solid samples by electroplating of elementary substances on the cathode or their oxidation on the anode. This procedure permits the preparation of thin deposits of uniform thickness. The radioactive substance is thus deposited in elementary form and sources may be prepared that have a maximum specific activity. This method is used with preference in the preparation of thin sources when it is desirable to exclude, or at least reduce self-absorption.

Standard electrolytic apparatus may be used in the preparation of samples by this method. The electrodes are standard plates of platinum, nickel, copper or stainless steel. They may sometimes be of spiral wire into which is inserted the counting tube during measurement. Electrode decontamination is rendered more difficult by radionuclide diffusion into the metal and therefore it is carried out immediately after measurement is completed.

Preparation of Solid Samples from Powdered and Gaseous Materials

Sometimes we are faced with the task of preparing a sample from radioactive powders. In the simplest case, the radioactive substance is spread in the measuring dish and its compactness is increased by gentle pressure on the dish or by means of a vibrator. With this manner of sample preparation particular care must be taken that all samples have at least approximately the same surface size and uniform layer depth. This is rather difficult with small amounts and hence the method is more suitable when larger amounts of radioactive powder are available. To improve mechanical strength, the surface may be covered by thin aluminum or plastic foil. In practice, various bonding agents are likewise used to this end [92]. The samples can be sometimes briquetted [93], or baked in an oven [88].

If, however, adequately stable samples are difficult to prepare directly from powders, it will be found more convenient to dissolve the powder in a suitable solvent and prepare the sample either by evaporation or by precipitation.

Solid samples may be prepared relatively readily even from some radioactive gases which form precipitates during absorption in an appropriate solution. For example, $^{14}CO_2$ is absorbed in a barium hydroxide solution and $Ba^{14}CO_3$ is precipitated. Further radioactive gases – $^{35}SO_2$, $H_2{}^{35}S$, etc. – may be processed and prepared similarly.

4.8.3 Liquid Samples

Preparation of Liquid Samples as an External Source

There are several instances where it is more convenient to carry out activity measurements with liquid samples. Some radioactive samples are originally liquid; others may be prepared by dissolution or melting of a solid substance, or dissolution of a radioactive gas. Measurement on liquid samples is justified mainly in the case of non-volatile or slightly volatile substances.

One advantage of activity measurements in liquid samples is that homogeneous and reproducible specimens may be easily and readily prepared. On the other hand, beta radiation is strongly absorbed in the liquid layer and self-absorption is therefore substantially greater than with solid samples. In view of this factor, it will prove more useful to measure the liquid samples in thickness greater than that of the range of the given beta radiation.

There are several ways of measuring liquid samples. The simplest of them involves placing the counting tube over the free surface of the liquid poured on a flat dish. The use of a limiting colimator is recommended: it is placed as near the surface of the sample as possible in order to rule out the effect of the meniscus. If the sample is extremely small and has a high surface tension, it will tend to form drops. This may introduce serious errors into the results because of non-defined or inadequately defined geometry, self-absorption and backscattering. A uniform distribution of the liquid at the bottom of the dish will be greatly aided by washing the dish carefully, and particularly by the addition of wetting agents. It is an indispensable condition not only that the volumes be identical but also that they always cover the same area. Otherwise, corrections, which are relatively complicated, but not too accurate, should be applied. One way to obtain a controllable area is to use dishes with concentric grooves in the bottom similar to those employed in the evaporation technique. The liquid fills up the smallest ring first and any excess passes on to the next. This permits relatively good areal distribution of small amounts, up to about 1 ml. Spotting on filter paper and similar absorbing material will sometimes prove useful [88]. Evaporation of volatile substances may be prevented by covering the surface of the sample by a thin film.

The radioactivity of liquid substances available in larger quantities may be measured by means of a special liquid counter. In such cases the radioactive solution comes into direct contact with the tube walls, which must be decontaminated immediately after the measurement. Gamma radioactive liquid samples may be conveniently measured by means of well-type scintillation crystals.

The results of measurements may also be affected by the density of the liquid. Hence, the densities of a whole series of specimens for measurement must be approximately equal.

Sample Mounting for Liquid Scintillation Counters

Samples to be measured by liquid scintillation counters are prepared by mixing a known amount of the radioactive liquid with scintillator solution. Only a few substances are capable of fulfilling the function of a scintillator, just as only a few solvents comply with the conditions for transmitting absorbed energy on to the dissolved scintillator. The most widely used scintillators and solvents are listed in Chapt. 3.1.2. The most widely used solvent is toluene. The presence of certain substances in the solution may have an inhibiting effect on the transmission of the absorbed energy. On the other hand, numerous agents significantly reduce the intensity of the light flashed produced in the interaction of beta particles with the liquid scintillator, i.e., they act as quenching agents. As radioactive substances become one of the components of the resultant scintillation sample, their properties must meet up to certain requirements. One of the basic properties of radioactive substances whose activity is to be measured by means of liquid scintillation counting is that they be soluble in a given solvent, or that a good contact be assured in some other way with the scintillator solution, and that they do not act as quenchers.

The simplest case is when the substance is soluble in the solvent utilized. This may be often attained by judicious selection of a suitable solvent for the given substance. If the latter is not dissolved in a simple solvent, e.g., in toluene, dioxane, etc., a suitable mixture of solvents is used. Thus, for radioactive substances present in an aqueous solution, a suitable solvent will be given, e.g., by a mixture of toluene, methylcellosolve and naphthalene, dioxane, etc. When dealing with aqueous solutions, it should be remembered that the dissolved radioactive agent may precipitate after mixing with the liquid scintillator.

Account must be taken of a further effect; the decrease of solubility with decreasing temperature. The sample at the detector head is usually maintained during measurement at a low temperature in a refrigerator and this may cause the radioactive substance to separate in solid form.

Another possibility is conversion of the radioactive substance to a suitable derivative soluble in toluene. Many metal cations can be converted to a chelate complex, which is soluble in toluene. Some important chelates have been mentioned in Chapt. 4.7.6. Sometimes the given cation may be transferred directly by extraction from the aqueous solution to toluene. Acids and bases may be converted to salts soluble in toluene by the addition of lipoidal bases or acids. This is also a suitable pathway for degradation products of hydrolyzing substances.

Particularly significant among organic bases is the high-molecular-weight quarternary ammonium hydroxide, called hyamine, and the high-molecular-weight aliphatic primary amine, known as primene. Hyamine is a very strong base and consequently forms salts even with very weak acids. Both are easily dissolved in toluene, xylene, water and alcohols. The activity of such acids as carboxyl and inorganic acids [94, 95], protein acids, amino acids, acid polysaccharides and CO_2 [96, 97], which are insoluble in toluene (like their normal salts), may be measured by the scintillation method in that solvent when they are added in the form of alcohol solutions of their hyamine or primene salts. Dialkylhydrogenphosphates and 2-ethylhexane acid may be considered as examples of organic acids which are capable of mediating the dissolution of bases, chiefly of metal cations in a scintillator solvent [98].

The problem presented by insoluble radioactive substances may sometimes be resolved by their dispersion in the solution in the form of a fine suspension whose stability becomes increased by the addition of a gel-forming agent. Aluminum-stearate and polymethylmethacrylate may serve as such; however, their drawback is that they are soluble only in hot toluene and on cooling form a stable gel system. Cab-O-Sil and Thixcin find a more general application here, as they do not require heating and may be added directly into the liquid scintillator, to which is also added the powdered radioactive substance or its aqueous suspension. Satisfactory dispersion is achieved by shaking the vessel, and measurement is carried out by the liquid-scintillation technique. Gel-scintillator, based on dioxane and silicon dioxide [99, 100] also has advantageous properties.

The activity of toluene-insoluble substances may be measured also by spotting their solutions on a filter paper which on drying is immersed into the vessel with the liquid scintillator. This method is similar to the suspension method which makes use of paper instead of a gel-forming agent [98]. This modification of the liquid scintillation method is of particular significance for chromatography and electrophoresis in which the spots are cut out and measured directly.

Radioactive substances that are added to the liquid scintillator should not be strong quenching agents. The quenching effect is brought about by interference with the emitted light flashes, or by their absorption. Typical quenching agents are mercaptans, phenols, amines, aldehydes, ketons, and nitro- and iodocompounds [50, 101].

Many radioactive substances act as quenchers if present in higher concentrations, but a certain degree of quenching is tolerable. Although a certain measure of absorption quenching always takes place in suspensions, it may nevertheless completely prevent measurement of colored samples. If the quenching is not too effective, a correction may be applied. Another possibility is the destruction or removal of colored substances. This may be done by means of oxidation, reduction, adsorption and other approaches. Certain substances that strongly quench in the solution may be conveniently measured in the form of suspensions. Sometimes it will be found more practical to convert strongly quenching agents by ashing to ^3H_2O or $^{14}CO_2$, absorbing these in a suitable agent, and then carrying out the measurements.

For useful hints on liquid scintillation techniques see also Ref. [102 and 103].

4.8.4 Gas Samples

Activity measurements of gaseous samples are met with less frequently than those of solids or liquids. Nevertheless, there are cases when measurement of activities in the gaseous phase cannot be avoided. Such is measurement of alpha activities in emanations, beta activity of ^{85}Kr, and generally, activities of inert gases. Exceptionally, it may be necessary to measure the activity of other gaseous elements, e.g., (^{15}O, ^{19}O) or nitrogen (^{13}N, ^{16}N).

On the other hand, preference is given to measurements of activities of ^{3}H, ^{14}C and ^{35}S in the gaseous state. These nuclides emit very soft beta radiation and therefore give rise to certain problems when measured as solid or liquid samples. Self-absorption in these samples (see Chapt. 4.8.2 and 4.8.3) brings about a marked decrease in measurement sensitivity and detection efficiency. If the sample is mounted outside of the detector, the sensitivity and efficiency are further reduced because of radiation absorption in the air and the counter window (protective foil in scintillation counters). The latter factor may be excluded by placing the sample directly into the sensitive volume of the detector. The effect of self-absorption, however, cannot be ruled out in solid and liquid samples and in the case of tritium renders detection by means of the ordinary technique practically impossible. The efficiency of measurement may range between 0.2 and 2.5 % [104, 105]. When measurement of such soft beta sources is involved, the method of gas measurement, which eliminates the effects of both absorption and self-absorption, may be conveniently used, besides that of liquid scintillators. The measured gas is one of the components of the gas filling in the ionization chambers, or proportional or GM counters.

The most important task in the preparation of gas samples for activity measurement is to assure purity of the radioactive gas and its transfer into the detector, where the vacuum technique is utilized.

Tritium and carbon-14 are very important in organic analysis. These radioactive elements are converted into the corresponding gas by ashing. Substances containing ^{14}C are converted to $^{14}CO_2$, or sometimes to acethylene, and measured as such. Tritium may be converted to various products, including water. When proportional counters are used for detection, foreign gases are generally added to the radioactive gas, e.g., methane, a methane–argon mixture, propane etc. Nevertheless, a pure $^{14}CO_2$ filling is also used [106, 107]. Pure carbon dioxide is inadequate to fill GM counters. Satisfactory discharge and quenching conditions may be assured by a mixture of CO_2 and CS_2. Both these counters, but especially the Geiger-Müller type, are extremely sensitive to impurities (e.g., O_2 or humidity), thus requiring careful purification before measurement. Ionization chambers are less sensitive to impurities.

^{14}C-labeled organic substances may be ashed according to one of the several known methods of dry [108–113] and wet [114–118] ashing. Some of these are also suitable for ashing of ^{3}H-containing compounds, for which, however, further procedures have been worked out [119–121].

During ordinary ashing processes, tritium is obtained in the form of T_2O. Its vapors do not lend themselves as detector filling for direct measurement, though this direct method may essentially be employed [122, 123]. Hence, direct measurement of water vapors is for the most part omitted; tritium is preferably converted to a suitable gas. Tritium oxide may be reduced to tritium by a reaction with magnesium [124], magnesium amalgam [125], calcium [126], $LiAlH_4$ [120], zinc [127] and sodium [125]. Nor is hydrogen the most suitable filling for a detector. It is better to convert T_2O to hydrocarbons such as methane, ethane, propane, etc., by the use of Grignard's reaction [112, 128], e.g.,

$$CH_3MgI + H_2O \rightarrow CH_4 + Mg(OH)I. \tag{4.24}$$

Also calcium carbide [129]

$$2T_2O + CaC_2 \rightarrow C_2T_2 + Ca(OT)_2, \tag{4.25}$$

or aluminum carbide [119]

$$Al_4C_3 + 12T_2O \rightarrow 4Al(OT)_3 + 3\,CT_4 \tag{4.26}$$

is used. In the above methods only half of the tritium is converted to hydrocarbon, which naturally reduces the efficiency of measurement.

A far more effective method is to convert the organic substance by Zn and NiO to a tritium methane and hydrogen mixture [121] which can be detected directly.

It will sometimes be found of advantage to measure ^{35}S activity in the form of an $H_2{}^{35}S$ and $^{35}SO_2$ mixture [130, 131].

4.9 Problems of Dosimetry and Radiological Safety

4.9.1 Introduction

All radioactive elements emit ionizing radiation (X-rays, gamma rays, alpha and beta particles). This radiation, like certain other types of radiation (heavily charged particles, neutrons), can interact with living cells and induce changes in their normal functions which may result in a whole series of harmful effects on the organism. A systematic review of these effects may be found in Ref. [132 and 133]. This, however, does not permit the conclusion that work with sources of ionizing radiation is dangerous and hence should be avoided. Work with radioactive materials and other sources of ionizing radiation presumes the introduction of safety measures to be one of the basic requirements; if such measures are strictly observed, then work with these materials is no more dangerous than any other activity in a modern laboratory. Therefore, work with sources of ionizing radiation need not inspire fear, but does require that the chemist be familiar with all the degrees of safety measures that must inevitably be observed. These are very simple not only in manipulations with many sealed sources, but also with small amounts of radioactive substances used as open sources in indicator methods.

From the point of view of safety in handling radioactive materials and other sources of ionizing radiation, it is important to avoid the possibility of external irradiation on any part of the body, or internal contamination of the organism with radioactive materials that would cause internal irradiation. Protection then consists in observing certain precautions that prevent such levels of external and internal exposure as might cause damage in the living organism.

4.9.2 Dose Units

A basic radiological quantity is a radiation dose which can be measured and expressed in several units. An ideal unit for measuring radiation doses from the viewpoint of radiological safety would be a unit that would express an equal biological effect regardless of the nature of the radiation and its energy. Unfortunately no such unit may be established in view of the complex biological effects of radiation. We must therefore be satisfied with units that are only an approximation. Essentially the following three units are employed.

The *roentgen* (r)—Initially, this unit was applied only to soft X-radiation. It was later extended to include X and gamma radiations with an energy up to 3 MeV. The roentgen is defined as "that exposure dose of X or gamma radiation such that the associated corpuscular emission per 0.001 293 g of air produces, in air, ions carrying 1 esu of quantity of electricity of either sign." One esu involves the production of 2.093×10^9 ion pairs, since the charge on an electron is 4.8×10^{-10} esu. The effective atomic number of the air is approximately equal to that of biological tissues and therefore, the number of produced ions, or of absorbed energy per unit of weight, is about equal, too. The applicability of this unit, is, however, markedly limited. Before all else, it may be used only for X and gamma radiation with an energy of up to about 3 MeV and is not able to provide a basis for the dosimetry of electromagnetic radiation of a higher energy, or dosimetry of corpuscular radiation.

The *rad*—This unit has a wider applicability, for it expresses the absorbed energy in a certain material regardless of the type and energy of radiation. The rad is defined as the absorption of 100 ergs g^{-1}. This absorbed quantity always refers to a definite material, in our case to 1 g of tissue. A dose that represents 1 *rad* in the air corresponds to an absorbed dose very close to 1 *rad* in normal body tissue. Consequently, in radiation protection, exposure in the air expressed in roentgens may be numerically taken as absorbed doses in tissues, in *rads*.

The *rem* (Roentgen Equivalent Man)—Equal doses of various types of ionizing radiation do not produce the same biological effects. Heavy charged particles, as for instance, heavy nuclei and alpha particles, have a higher biological effectiveness per given absorbed amount of energy than gamma or beta radiations. It is evident that these differences must be expressed from the aspect of radiation protection. The unit for a biologically effective dose (also designated as an equivalent dose), recommended by the International Commission on Radiological Protection (I.C.R.P.), is 1 *rem*.

Originally 1 *rem* was defined as a dose which produces the same biological effects as 1 roentgen of X or gamma radiation. Because the *rad* has come recently to be used as a unit of physical dose, the unit of biologically effective dose is also referred to this physical unit. To differentiate between the various biological effects of particular types of radiation, we use factors of relative biological effectiveness (RBE). The value of this factor is defined as the ratio of the doses of a certain reference radiation to the particular radiation which produces an equal biological effect. X-radiation (200–250 keV) is used as reference. The value of the biologically effective dose D_{rem} is then obtained by multiplying the physical dose D_{rad} by the coefficient of the biological efficiency

$$D_{rem} = (\text{RBE}) \, . \, D_{rad} \, . \tag{4.27}$$

The following RBE factors are used: RBE = 1 for X, gamma, beta positron and negatron radiation; 2.5 for thermal neutrons; 10 for fast neutrons; 10 for alpha particles and protons; 20 for heavy ions.

4.9.3 Tolerance Dose

Much study has been needed to determine the value of a tolerance dose. One of the difficulties is the determination of ionizing radiation dose, for only the dose received at the work-place is checked and recorded. Another complication arises from the fact that many of the effects become manifest very slowly, after many years. Nevertheless, the results of intensive studies in the post-war years and a pooling of concepts and experiences in the field of radiobiology have made it possible to set up a tolerance dose. If it is inexact, the consequent hazard is so small that the ensuing risk somatic damage and genetic effects may be neglected in comparison with other risks in life. Tolerance doses ought to be viewed as maximum values which we strive never to reach. It should further be observed that the tolerance doses stated here are meant for professionals working with radiation sources and not for the population as a whole.

In the majority of countries where radionuclides and other radiation sources are widely applied, the system of radiation protection is based on the I.C.R.P. recommendations [134–138]. Tolerance doses are expressed in *rem*, which is of advantage in that various types of ionizing radiation may, with the help of the RBE factors, be set up in a single system. This concept of relative biological effectiveness is of significance only when other factors are taken into account, e.g., the nature of the relevant organ of the body, its physiological state, the type of the observed radiation, dose intensity, etc. In connection with the tolerance doses and radiation protection, radiation is differentiated into external and internal.

External Radiation

The permissible total-body dose of external radiation has been determined on the basis of the permissible dose for the most sensitive organs of the human body. These are the gonads, where damage is brought about by mutations, and other areas of possible immediate somatic injury: the blood-forming organs and the crystalline lens. In view of the possibilities of these injuries, it is recommended that the maximum permissible dose for persons over 18 years of age, occupationally working with ionizing radiation, be expressed according to the relation

$$D_{\text{rem}} = 5(N - 18), \tag{4.28}$$

where D_{rem} is the dose in *rem*, and N the person's age.

This implies that the allowable yearly dose for persons above the age of 18, is 5 *rem* from which an overall permissible weekly dose of 0.1 *rem* has been determined. This value is a directive for planning and projecting. Often, the dose is made to refer to a 13-week period during which the maximum permissible dose for the most sensitive body organs is 3 *rem* on condition that it will not exceed 5 *rem* within a year. Moreover, 3 *rem* represents the highest possible dose that may be received in a single exposure, if a corresponding reduction in exposure is assured in the following time-period, together with a balancing of the total permissible dose. Higher single doses must be ruled out. An important condition for the validity of relation (4.28) is that the dose is smoothly absorbed.

In the human body, however, there are less sensitive organs where higher doses are allowable. A review of the maximum permissible doses for various parts of the body are shown in Table 4.4.

Internal Radiation

Internal exposures are those in which radioactive material has entered the body by ingestion, by inhalation or through the skin. If a radioactive element enters the body, it reaches various organs by pathways that depend primarily on the chemical nature of the element. It may remain for a very long time, during the whole life of

Table 4.4

Maximum Permissible Radiation Doses for Various Parts of the Body

Part of body	Maximum permissible dose (rem)	
	in 13 weeks	in 1 year*
Gonads, blood-forming organs, crystal. lens	3	5
Hands, forearms, soles, ankles	20	75
Remaining parts of the body	8	30

* Relation (4.28) may be applied.

the organism, and be permanently active at one site if its half-life is sufficiently long, Therefore, internal contamination may entail grave consequences even if a relatively small amount of radioactive elements is involved. Radioactive elements may be inhaled in the form of vapors or dust (respiratory system), may be ingested with food or drink (alimentary tract), or may enter through wounds or be directly absorbed through the skin.

Particular radionuclides differ in radiotoxity and hence do not present the same dangers. Their toxicity depends in particular on their half-life, the type and energy of the radiation, metabolism of the element and the biological excretion rate of the radionuclide from the organism. In general the more toxic a radionuclide, the longer its half-life, the greater the ionizing power of its radiation, the more selectively it is stored in certain parts of the body and the more difficult it is to remove.

The measure of the rate of excretion of a radioactive element from the body is the so-called *biological half-life* (T_B). This is the time required for the radioactive element's concentration in the body, by virtue of biological processes, to be reduced to its half. The biological and physical half-lives are important criteria for judging radiotoxicity and both are expressed in terms of a common value by the so-called *effective half-life* T_{Ef}

$$T_{Ef} = \frac{T \, . \, T_B}{T + T_B} \, . \qquad (4.29)$$

The effective half-life may substantially differ from the physical half-life and is the chief measure of radiotoxicity.

Alpha and beta emitters are dangerous from the point of view of internal contamination and radiation, while they do not represent any particular hazard as regards external irradiation. Internal irradiation by alpha sources is the more serious, as its factor of relative biological effectiveness, RBE = 10, contrasts to beta and gamma

Table 4.5

Permissible Amounts of Some Radionuclides in View of Their Internal Action

Radionuclide		Permissible dose in the body (μCi)	Maximum permissible concentration during a 40-hr working week (μCi cm^{-3})	
			Water	Air
^3H ^3H_2O		10^3	0.1	2×10^{-5}
^{14}C $^{14}CO_2$		300	0.02	4×10^{-6}
^{24}Na	s	7	6×10^{-3}	10^{-6}
	is		8×10^{-4}	10^{-7}
^{32}P	s	6	5×10^{-4}	7×10^{-8}
	is		7×10^{-4}	8×10^{-8}
^{55}Fe	s	10^3	0.02	9×10^{-7}
	is		0.07	10^{-6}
^{59}Fe	s	20	2×10^{-3}	10^{-7}
	is		2×10^{-3}	5×10^{-8}
^{60}Co	s	10	10^{-3}	3×10^{-7}
	is		10^{-3}	9×10^{-9}
^{89}Sr	s	4	3×10^{-4}	3×10^{-8}
	is		8×10^{-4}	4×10^{-8}
^{90}Sr	s	2	4×10^{-6}	3×10^{-10}
	is		10^{-3}	5×10^{-9}
^{131}I	s	0.7	6×10^{-5}	9×10^{-9}
	is		2×10^{-3}	3×10^{-7}
^{210}Pb	s	0.4	4×10^{-6}	10^{-10}
	is		5×10^{-3}	2×10^{-10}
^{210}Po	s	0.03	2×10^{-5}	5×10^{-10}
	is		8×10^{-4}	2×10^{-10}

s — soluble form, is — indicates insoluble form.

sources where the RBE = 1. Beta sources are more dangerous than gamma emitters despite their RBE factors being equal: beta radiation loses all its energy within the body, or even within the affected organ itself, while gamma radiation does not.

The most dangerous are elements which are selectively deposited in the bones, the liver, and other essential organs and which, moreover, have a high effective half-life. An accumulation of a radioactive element in any one organ leads to its rapid impairment. Thus, for example, alkaline earths (calcium, strontium) are deposited in the bones, and yttrium, zirkonium and polonium in the bone marrow, the liver, etc.

The site of action is often substantially affected by the half-life of the radioelement. For instance, radioactive phosphorus passes in the form of phosphate into the soft tissues first, and only after several days or even weeks when its activity has significantly decreased is it bound to calcium in the bones. The site and duration of a radioactive element sometimes depend on the chemical form in which it reaches the organism. Radioactive carbon, for instance, present in inorganic compounds is rapidly excreted as carbon dioxide and is harmless. If, however, it enters the body in the form of organic compounds, it must pass through complex metabolic processes and takes substantially longer to be excreted from the organism. Sometimes even the manner in which a radioactive element enters the body (inhaled, ingested, through skin), may be of significance.

All the above factors have to be taken into account when permissible doses for individual radioactive elements in the body, and also their permissible concentrations in air and drinking water are to be determined. Data on the tolerated quantities of individual radionuclides in the body and their tolerated concentrations in the air and drinking water, as adopted by the I.C.R.P., may be found in relevant reports [134, 138–140], together with the methods of determining permissible doses in the presence of several radionuclides. A short extract from the tables in these sources, for certain important radionuclides, is given in Table 4.5. The values stated therein apply even to permanent incorporation, on condition that no important external irradiation of the subject takes place in addition to the internal effects. If the subject has been externally irradiated as well, then the listed values must be adjusted downward so that the combined dosage does not exceed that permitted for each organ.

4.9.4 Protection from External and Internal Irradiation

4.9.4.1 General Principles of Protection

To assure safety during handling of radioactive materials it is necessary to prevent the possibility of the organism's being irradiated from both internal and external sources, and this fundamental condition ought to be kept in mind by everyone professionally engaged in work with radioactive substances. Although the degree of the indispensable safety measures differs in every case, nevertheless certain general rules of protection against both internal and external contamination may be set down.

In work with gamma radioactive substances, whether the sources are open, sealed, or in some other form, such as neutron radiation, the danger of external irradiation prevails. In consequence, measures are taken to reduce dose intensities (i.e., doses received by environment per unit of time) or doses not only to the permissible level, but if possible to a level close to background values.

The following relation is valid for a dose of gamma radiations in rads, at a distance d from the point source (in meters) for time t (in hours):

$$D_{rad} = \frac{0.5 \,.\, A \,.\, E \,.\, t}{d^2}, \tag{4.20}$$

where A indicates the activity of the source of radioactive radiation, in curies, and E the energy of gamma radiation, in MeV. This relation implies that the value of the radiation dose of a given source is directly proportional to the time duration of irradiation and indirectly proportional to the square of the distance from the radiation source. Accordingly, the radiation dose may be reduced either by shortening the duration of irradiation, or by increasing the distance from the source. The first condition, shortening of the time during which a person is exposed to radioactive radiation, is achieved by the use of rapid working procedures and a rational work organization. Increasing the distance is made possible by the use of special equipment, working tools and remote control manipulators.

However, these two methods will not suffice in the majority of cases to keep the dose intensity of radiation at the required level. Hence, a further means is employed: shielding the worker from the radiation source by interposing protective walls and screens of suitable thicknesses and material to absorb the emitted radiation and thereby lower the dose intensity. The possibilities of screening, in view of the activities involved, are practically limitless, and their applicability is discussed in the next chapter.

When radioactive substances are manipulated as open radiation source, attention must be devoted in particular to protection against internal irradiation, i.e., preventing radioactive elements from entering the organism. This may be achieved, above all, by: (1) excluding the possibility of a direct contact between the organism and the radioactive elements, (2) preventing radioactive gases, vapors or powders from entering into laboratory air, (3) preventing or at least reducing to a minimum the possibility of radioactive contamination of the laboratory, and (4) providing a light decontamination. Implementation of these conditions is assured by correct location, suitable equipment, ventilation of laboratories, individual protective devices, special working methods and, finally, by a sound organization of the work in question.

4.9.4.2 Shielding

Beta Radiation Shielding

Beta radiation sources as external sources represent only a minor hazard. The intensity of beta radiation may be substantially reduced by the air itself. The maximum range of beta radiation with energies between 0.5 and 3 MeV is between 1 and 10 m. In the case of other materials this corresponds to relatively thin layers. The gradual loss of energy in its passage through a substance is accompanied by the formation of electromagnetic radiation, the so-called "Bremsstrahlung". This is far more penetrating than the original beta radiation and for effective shielding against the beta particles it is advisable to reduce its formation to a minimum. The mean energy loss by beta particles through the production of "Bremsstrahlung" is approximately proportional to Z^2 of the absorbing medium, and increases almost linearly with increasing energy. Hence, it is applied above all for shielding beta sources with a higher energy and a high atomic number. Consequently, for beta radiation we choose shielding materials with low atomic numbers, such as plastics, aluminum, glass, and so on. When soft beta radiation sources are to be shielded the effect of "Bremsstrahlung" may be neglected. In the case of hard beta radiation it is advisable to shield the source first with a light material that will absorb the primary radiation with a minimum production of "Bremsstrahlung," which is then shielded further with a heavy material suitable for gamma radiation. When positron radiation is involved, account must be taken also of annihilation radiation which affects the calculation of the secondary shielding. The diagrams in Figure 4.27 enable the thicknesses of various absorbing materials to be determined, to assure full absorption of beta radiation with a given maximum energy.

Gamma Radiation Shielding

Shielding of gamma radiation is far more complicated, for this radiation may lose its energy through photoeffect, Compton scattering and electron pair formation. The relative occurrence of these three processes depends on the energy of gamma radiation and the nature of the absorbing material.

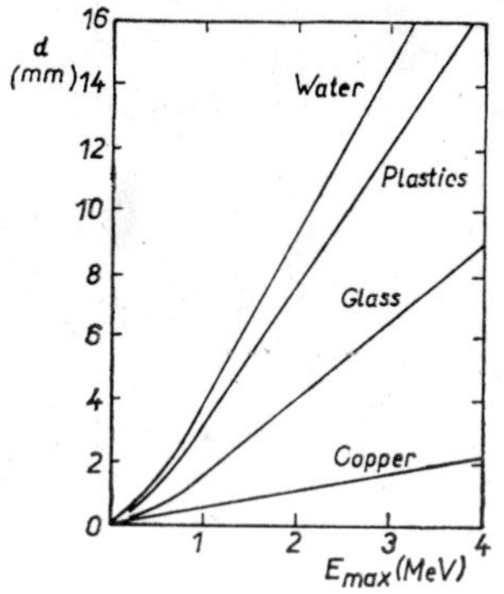

Fig. 4.27. Absorber thicknesses d for complete screening of a beta radiation source of given maximum energy E_{max}.

An important point for the calculation of screening is differentiation of the geometrical conditions and in principle the geometry of a "narrow" and a "broad" beam. In the first case a simple exponential law and the use of current diagrams of dependence, e.g., absorption coefficient–energy; half absorption thickness–energy, etc., as stated under 4.5.3.3, may suffice for a calculation of attenuation. When the broad beam is involved the ratio of the transmitted radiation I/I_0 is greater, by the produced and detected scattered radiation, than that for the same thickness with a narrow beam as is evident from Fig. 4.28. This requires a correspondingly greater shielding thickness.

The most important and the most widely used factors in work with radioactive gamma nuclides are the so-called *build up factors* (B-factors). For a dose intensity, B-factor is defined as the ratio of the true dose intensity of the observed gamma radiation, under conditions of a broad beam at a certain point within or close behind the shielding layer of a given thickness, to the dose intensity that would be calculated according to a simple exponential law for a narrow beam; it is designated as B_D. B-factors may relate also to energy (B_E), or the number of photons (B_N).

The value of B-factors depends not only on the energy of the primary radiation, absorber thickness and its atomic number (in the case of multi-component systems –on the effective atomic number), but also on the geometric conditions of the source, the shielding material and detector. The significance of B-factors is evident from Fig. 4.28. B_D-factors are tabulated or plotted on graphs, see Ref. [142].

On the basis of correlations between the relative ratio of transmitted radiation I/I_0

and the thickness of the shielding layer, the graphical correlations shown in Fig. 4.29 are given for various energies and materials. They enable us to determine the thicknesses of the shielding layer that gradually attenuate the value of the dose intensity by a factor of 10.

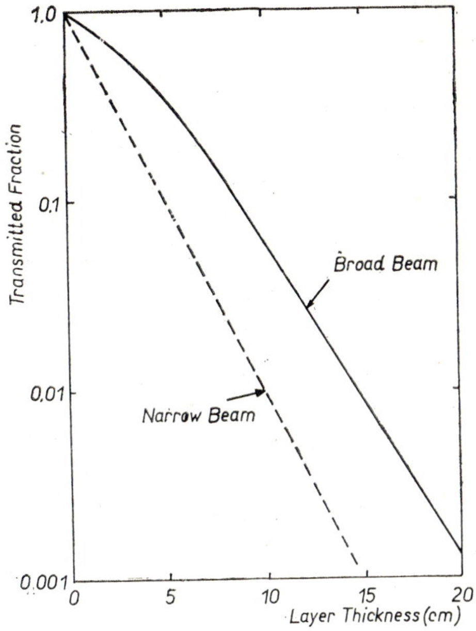

Fig. 4.28. Gamma radiation transmission of energy 1.0 MeV through a layer of iron, in a narrow and a wide beam.

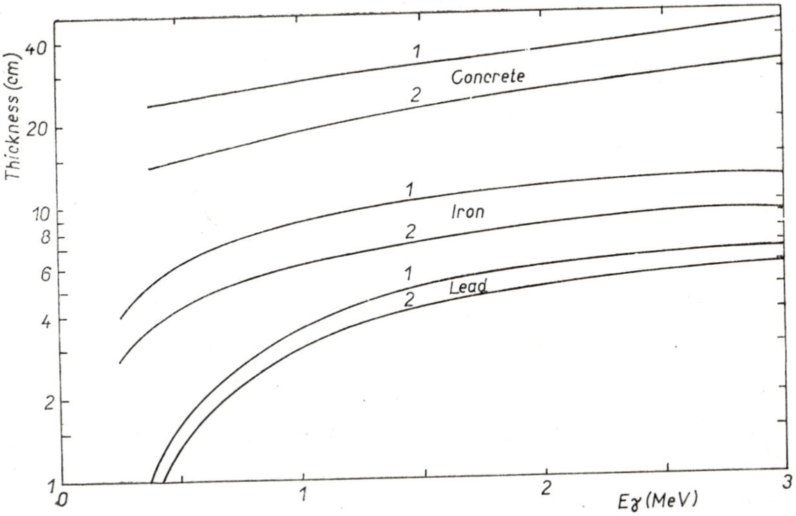

Fig. 4.29. Screening thickness required for a tenfold reduction of the dose intensity for gamma radiation (1 — thickness needed to reduce the original radiation by a factor of 10, 2 — thickness needed for the next reduction by a factor of 10).

It may likewise be found expedient to express the dose intensity in maximally admissible levels D_r, by means of the relation [143]

$$D_r = B_D \frac{fA\varrho}{4\pi d^2 \varepsilon} e^{-\mu R}, \tag{4.31}$$

where A is the activity of the source in disintegrations per second, f the number of photons per disintegration, ϱ the density of the shielding medium, R the absorber thickness (product of $d . \varrho$), ε the number of photons corresponding to the admissible dose intensity (depends on photon energy, approximately $\varepsilon = 1300/E$), and μ the absorption coefficient of the initial gamma radiation.

When determining the thickness of the shielding layer we may make use of the tables of the attenuation coefficients K that are listed for various types of shielding materials. The values in the tables were calculated on the basis of the attenuation of a broad beam and the increase of the dose intensity after shielding under the effect of a multiple scatter of gamma radiation [144]. Hence, the starting data in the tables are not based on the activity of the source but on the coefficient of the required attenuation. This enables the tables to be applied universally; it is however, of importance to determine beforehand the dose intensity of the unscreened source, which, with the exception of the point source, is also a rather laborious operation. Fortunately, the majority of laboratory radiation sources used for analytical purposes, admit an approximation of the point source. An extract from Gusev's tables [144] is presented in Table 4.6.

Radionuclides emitting gamma radiation of various energies require that calculations be made for each energy separately and the results combined in such a manner as to enable the share of each energy component of gamma radiation in the total dose to be taken into account. In the case of current gamma sources this may be done according to West [150].

In addition to these references to gamma radiation shielding already cited, Ref. [145–149] provide further details. It is advisable always to measure, as far as possible, the dose intensity after shielding—even during the experiments. Materials commonly used for shielding are chiefly lead, steel and concrete.

Neutron Shielding

Protection against fast neutrons depends on their slowing down during their passage through the material containing light atoms; these slow neutrons are then absorbed in substances with a large cross-section for neutron capture. Numerous substances become radioactive under the effects of neutron radiation and this fact has to be taken into account when the material is being selected.

To slow down neutrons, substances containing a high percentage of hydrogen, e.g. paraffin, water, plastic materials, and concrete, are used. Boron and various boron-

Table 4.6

Thicknesses of Protective Walls against Gamma Radiation for Various Coefficients of Attenuation K for Concrete, Iron and Lead (in cm)

K	Energy of gamma radiation, MeV								
	0.1	0.2	0.4	0.6	1	1.5	2	3	6
	Concrete ($\varrho = 2.3$ g cm^{-3})								
2	4.7	7.6	11,3	12.4	12.9	13.6	14.1	15.3	18.8
5	5.6	11.0	18.8	21.8	23.5	25.8	28.2	32.9	38.7
10	8.2	14.6	23.7	26.8	29.9	34.0	37.7	43.4	51.6
50	9.9	18.8	30.8	37.6	44.6	52.1	58.1	66.9	81.6
10^2	11.5	21.1	35.2	43.0	50.5	58.3	65.7	77.5	95.1
10^3	15.5	28.8	48.1	59.2	70.4	81.7	92.7	110.9	137.9
10^4	18.8	35.2	60.3	74.7	89.2	104.5	118.6	143.2	179.0
10^5	30.5	50.5	75.1	88.3	106.8	126.6	144.4	173.8	218.4
	Iron ($\varrho = 7.89$ g cm^{-3})								
2	0.7	1.2	2.2	2.7	3.3	3.6	3.9	4.4	4.6
5	1.4	2.5	4.1	5.1	6.4	7.4	8.1	8.9	9.6
10	1.9	3.5	5.6	6.8	8.5	10.0	11.0	12.2	13.2
50	2.9	5.2	8.4	10.3	12.7	15.1	16.9	19.1	21.5
10^2	3.4	6.1	9.6	11.7	14.5	17.3	19.5	22.1	25.5
10^3	4.5	8.2	13.2	16.3	20.5	24.4	27.5	31.7	36.0
10^4	6.8	11.5	17.1	20.7	26.0	31.3	35.5	40.9	46.5
10^5	10.0	15.8	20.8	24.9	31.5	38.0	43.2	50.0	58.3
	Lead ($\varrho = 11.34$ g cm^{-3})								
2	0.1	0.2	0.4	0.7	1.3	1.7	2.0	2.1	1.6
5	0.2	0.4	0.9	1.5	2.8	3.8	4.3	4.6	3.8
10	0.3	0.55	1.3	2.1	3.8	5.1	5.9	6.5	5.5
50	0.4	0.85	1.95	3.25	6.0	8.2	9.6	10.6	9.2
10^2	0.5	1.0	2.3	3.85	7.0	9.65	11.3	12.2	10.9
10^3	0.7	1.5	3.3	5.7	10.2	14.1	16.5	18.0	16.5
10^4	1.05	2.1	4.55	7.5	13.3	18.3	21.3	23.5	22.0
10^5	1.15	2.4	5.4	9.2	16.5	22.7	26.2	28.9	27.5

containing substances (boral, boron-steel), cadmium and concrete are employed to capture slow neutrons.

Water is used not only as a moderator but also as a protective material. After several collisions with hydrogen atoms, a neutron slows down to thermal energy and is ab-

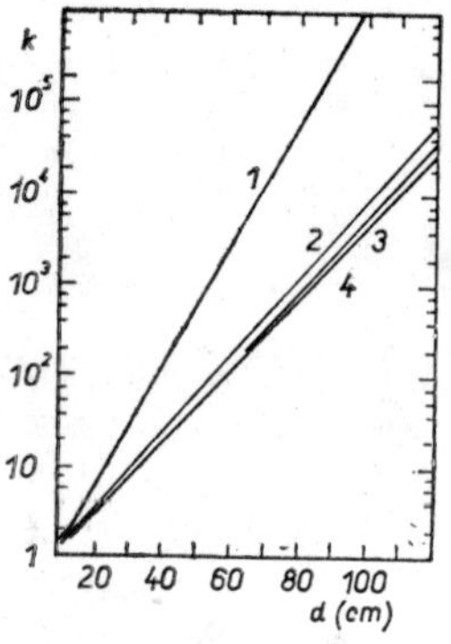

Fig. 4.30. Attenuation coefficient in dependence of the water layer thickness for neutron sources (1: Po–B, 2: Ra–Be, 3: Po–Be, 4: Pu–Be).

sorbed because of the H (n, γ) D reaction; gamma radiation of an energy of 2.23 MeV is emitted, which should be shielded too. The energy of gamma radiation due to neutron capture may be significantly reduced (down to 0.5 MeV) by absorption of thermal neutrons in boron. Cadmium is known for its high cross-section for neutron capture, with an energy lower than 0.6 MeV. A cadmium sheet 0.1 mm thick is capable of reducing the thermal neutron flux by a factor of 10^9. This, however, gives rise to gamma radiation with an energy of 7.5 MeV. Because of its better mechanical properties Cd – Pb alloy is also applied.

A relatively thin shielding layer is adequate for shielding laboratory neutron sources. Calculations may be carried out with the aid of nomograms set up from data of a space-energy distribution of neutrons in an infinite shielding material [151]. Figure 4.30 shows the relationship between the coefficient of attenuation and the thickness of the water shielding for various neutron sources.

For laboratory sources the accuracy of the calculation based on the following formula is sufficient.

$$\Phi = \frac{SB}{4\pi r^2} e^{-\Sigma_{\text{rem}} d}, \qquad (4.32)$$

where S is the total emission of the source into the spatial angle 4π sr, r the distance from the source in centimeters, constants B and $\sum_{\text{rem}}$ (macroscopic cross-section) characterize the given source and shielding material, and d the thickness of the neutron screening. For $d > 20$ cm in the case of paraffin it is recommended that we take $B = 5$. For sources Ra – Be and Po – Be we take $\sum_{\text{rem}} = 0.16$ for paraffin. Values $B\,e^{-\Sigma_{\text{rem}} d}$ $(= b_d)$ for various shielding thicknesses are given in Table 4.7.

Table 4.7

Values $Be^{-\Sigma_{rem}d}$ for Various Shielding Thicknesses with Paraffin

d (cm)	20	22	24	26	28	30	35	40	45	50
b_d	0.25	0.18	0.13	0.10	0.07	0.05	0.02	0.01	0.006	0.003

To calculate the dose intensity in maximally permissible levels, we may apply the formula

$$D_r = \frac{BS}{4\pi r^2 \varepsilon_n} e^{-\Sigma_{rem} d}, \tag{4.33}$$

where ε_n is the neutron flux of a given energy which causes the maximally permissible dose intensity.

For further details see Ref. [141, 151–153].

4.9.4.3 Equipping Laboratories for Work with Radioactive Substances

The equipment used in a laboratory will depend on the level of activity and radiotoxicity of the radionuclides that are processed in it. The International Atomic Energy Agency (I.A.E.A.) has recommended that radionuclides be classified according to their relative radiotoxicity [154], as shown in Table 4.8.

Laboratories with radioactive substances are further divided, according to the level of the activity of radionuclides of the above four groups of radiotoxicity, into three categories: low (I), medium (II) and high (III) activity level. Table 4.9 indicates maximum activity levels for radionuclides handled as open sources in the various categories of workplaces.

It is to be noted, however, that the upper limit in category II may vary depending on the standard of equipment of the workplace. The values shown in the table are to be considered as maximal. The degree of hazard in the handling of radioactive materials also depends on their physical state and on the procedures selected. In this connection coefficients are frequently employed that express the degree of difficulty.

A well-equipped chemical laboratory may without any rebuilding conveniently serve as workplace for low activities (category I). It should nevertheless be designated exclusively for work with radioactive substances, and not used simultaneously

Table 4.8

Classification of Radionuclides According to their Relative Radiotoxicity (I.A.E.A.)

Toxicity group	Radionuclides
A: extremely high	$^{90}Sr + {}^{90}Y$, $^{210}Pb + {}^{210}Bi$ (RaD + E), ^{210}Po, ^{211}At, ^{226}Ra, ^{227}Ac, ^{233}U, ^{239}Pu, ^{241}Am, ^{242}Cm
B: high	^{45}Ca, ^{59}Fe, ^{89}Sr, ^{91}Y, $^{106}Ru + {}^{106}Rh$, ^{131}I, $^{140}Ba + {}^{140}La$, $^{144}Ce + {}^{144}Pr$, ^{151}Sm, ^{154}Eu, ^{170}Tm, $^{234}Th + {}^{234}Pa$, natural thorium, natural uranium
C: medium	^{22}Na, ^{24}Na, ^{32}P, ^{35}S, ^{36}Cl, ^{42}K, ^{46}Sc, ^{47}Sc, ^{48}Sc, ^{48}V, ^{52}Mn, ^{54}Mn, ^{56}Mn, ^{55}Fe, ^{58}Co, ^{60}Co, ^{59}Ni, ^{64}Cu, ^{65}Zn, ^{72}Ga, ^{74}As, ^{76}As, ^{82}Br, ^{86}Rb, $^{95}Zr + {}^{95}Nb$, ^{95}Nb, ^{99}Mo, ^{96}Tc, ^{98}Tc, ^{105}Rh, $^{103}Pd + {}^{103}Rh$, ^{105}Ag, ^{111}Ag, $^{109}Cd + {}^{109}Ag$, ^{113}Sn, ^{127}Te, ^{129}Te, ^{132}I, $^{137}Cs + {}^{137}Ba$, ^{140}La, ^{143}Pr, ^{147}Pm, ^{166}Ho, ^{177}Lu, ^{182}Ta, ^{181}W, ^{183}Re, ^{190}Ir, ^{192}Ir, ^{191}Pt, ^{193}Pt, ^{196}Au, ^{198}Au, ^{199}Au, ^{200}Tl, ^{202}Tl, ^{204}Tl, ^{203}Pb
D: low	^{3}H, ^{7}Be, ^{14}Ca, ^{18}F, ^{51}Cr, ^{71}Ge, ^{201}Tl

Table 4.9

Maximum Activity Levels for Radionuclides in Laboratories of Various Categories

Toxicity group	Activity range (mCi)		
	I	II	III
A	up to 0.01	0.01–10	above 10
B	up to 0.1	0.1–100	above 100
C	up to 1	1–1000	above 1000
D	up to 10	10–10 000	above 10,000

for other laboratory procedures. Separate rooms should be available for storing radioactive substances, washing vessels and, as far as possible, measuring radioactivity. If this is not possible, then at least sufficient space facilities for work must be provided. The greatest cleanliness must be maintained in the laboratory and workplaces must be protected against contamination. Trays covered with absorbent material such as filter paper are used in every handling or working performance. Laboratory personnel must always use protective clothing (rubber gloves, overalls, shoe covers, etc.) and abide by all the safety regulations. A regular dosimetric check-up of the workplace and all personnel must be carried out at regular intervals.

Safety rules governing work in laboratories of the middle (category II) and high (category III) activity level are far more exacting. A well-equipped workplace for the medium level of activity – the majority of radiochemical workplaces – must comply with the following requirements: The entire working complex must be divided into active and nonactive sections, connected by the "hygienic loop" – consisting of an active and a nonactive cloak-room, between which are washing rooms with showers and controlling instruments. The active section must include at least two laboratories with different activity levels, a separate laboratory for an eventual neutron source and individual rooms for storage and preparation of solutions, decontamination, radioactivity measurements and possibly also for autoradiography. Laboratories and rooms for storing radioactive materials and preparing solutions must be equipped with fume hoods, dustproof cupboards and eventually with special manipulation tables. The greatest possible selection of remote control manipulators should be available together with materials for setting up shielding screens of diverse thicknesses. Such a workplace must be thoroughly equipped with for each individual means of protection and dosimetric instruments. Floors, walls, doors, furniture and all other equipment should be smooth and easily washable. The whole active section is connected with air-conditioning equipment assuring an adequate inlet of fresh air, made to stream solely in the direction of the higher activity. Fume hoods and dustproof cupboards are ventilated individually. The active department is provided with a special canalization system leading to a collecting sink in which the radioactive waste is suitably treated, diluted and removed. Purely chemical laboratories and the other nonactive workplaces are set up separately, but in the immediate vicinity of the active department.

Departments with a high activity level are set up only in special plants and institutes whose working activity also includes production of radionuclides or labeled compounds.

Special Laboratory Equipment

The better the special equipment and installations in radiochemical laboratories, the safer the work. Hence, we endeavour always to equip laboratories with the best instrumentation, and never allow work with radioactive substances to be carried out under conditions that might prevent essential safety precautions from being taken.

Furniture, for example, should reduce to a minimum the possibility of its becoming contaminated. It should be of simple shape, having a suitable coating that

Fig. 4.31. Glove box.

can be easily washed, and rest on legs. Working tables and consoles should comply with the same requirements. The working surface must be smooth, and impermeable to aqueous and organic solutions, and tables should be provided with drains along their perimeter for easy washing.

The most important installations in a radiochemical laboratory are fume hoods. Most of the radiochemical operations are carried out under them. They can be hermetically closed and have their own ventilation system which leads through a special exchangeable filter. A fume hood must be correctly located, for instance, away from sources of strong drafts (such as doors, windows, ventilation), for this might reduce the ventilating (air-conditioning) efficiency or even make possible contamination of the air and the room. The fume hood is constructed in such a manner and of such material as to permit easy washing and resist corrosion. Most frequently it is made of stainless steel of high luster. The working surface is molded in the form of a flat dish, the bottom of which is lined with absorbent material such as filter paper or asbestos; these are replaced when contaminated. The fume hood is suitably provided with water, gas, and electricity supply. Wall thickness is so selected as to ensure the necessary protection even against external irradiation. Fume hoods for processing higher activities have a mirror attached to the back wall in which handling and the course of operations behind the shielding wall can be observed. In such cases the fume hoods are adjusted for work with remote control manipulators.

When simpler operations are involved, especially in work with beta radiation sources and low activities of gamma sources, closed boxes (manipulation boxes) are employed. Their construction depends on the nature of the work for which they are intended. Accordingly their basic parts may be of stainless steel, or other materials which are then coated with a hard shiny paint. The window is commonly of glass or a transparent plastic substance. Sometimes the whole box is made of plastic material. The front wall carries openings for the hands, to which rubber gloves are attached. The boxes have, for the most part, side chambers through which required objects are inserted or removed without necessitating opening of the principal box. They are movable, otherwise resembling fume hoods. An illustration of a closed box is shown in Fig. 4.31.

Fume hoods and manipulation boxes provide protection against internal contamination through hermetical closure, and against external radiation through shielding walls and the provision of remote control manipulators. All operations that may give rise to the formation of gases, steam, dust or where there is danger of spilling a solution, are performed in these installations.

Special protective containers have been designed for transporting, transfering or short-term storage of radioactive materials. They are usually cylindrical and are made of lead or steel for gamma sources, and of aluminum, plastics or other light material for beta and alpha sources. Containers for various radionuclides are selected with wall thickness that will prevent the dose intensity at the outer surface from exceeding, the permissible level. For radiation shielding we use protective shields or walls built of various materials which are selected in accordance with the principles set down in Chapt. 4.9.4.2. Protective shields may be movable and their front wall may carry openings with rubber gloves for the hands, or openings for exchangeable manipulators.

In order to exclude immediate contact with radioactive materials or containers in which they are held during various operations, we use special working tools enabling handling at a safe distance. The simplest of them are grasping tongs; these are available in various designs according to the shape and size of the object or vessels and they are destined to grasp (Fig. 4.32), and are made of steel. To prevent glass vessels from breaking, the ends of the tongs are coated with a layer of rubber or plastic material.

The length of the arms depends on the distance required between the operator and the radiation source. In more complex cases when the operator's movements and action are to be transferred to greater distance, remote manipulators are used. The simpler ones are able to carry out a few simple operations, while the

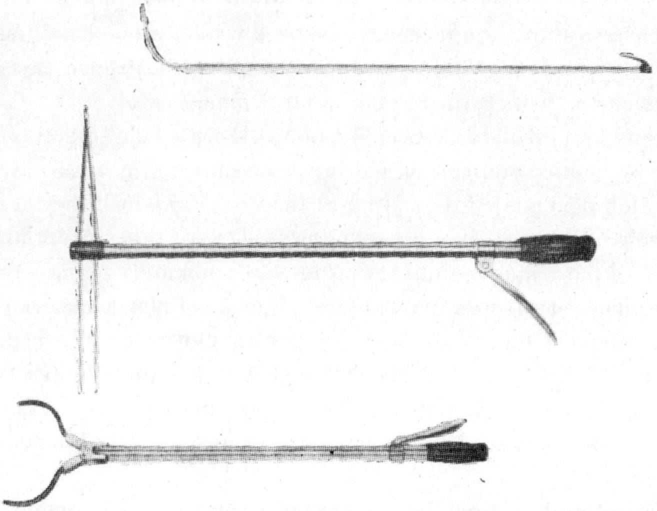

Fig. 4.32. Remote handling equipment with simple manipulators.

more sophisticated manipulators are capable of reproducing the movements of human hands and are versatile enough for complex working operations.

For work with radioactive substances we generally make use of ordinary chemical vessels, as far as possible those with the best properties. In view of the fact that these vessels are often handled by mechanical tongs and manipulators, they should possess high mechanical strength and occasionally also resistance to high temperatures and chemical agents. In contrast to glassware, silica vessels are noted for their lower adsorption ability and hence they are preferred for work with small concentrations of radioactive substances. In addition, the natural radioactivity of glass should be taken into account, arising from ^{40}K. Besides glass and silica used for vessels, trays, folia, sintered ware and various other objects, radiochemical laboratories, also make use of copper, china, aluminum, stainless steel, nickel, platinum, polyethylene, and others. The selection is determined in particular by resistance to temperature, coefficient of expansion, resistance against common chemical agents, and adsorption ability.

The basic equipment of a radiochemical laboratory comprises also instruments and aids for dosimetric controls, based on the principle of ionization chambers, counters and film dosimetry. Personal dosimetry includes film dosimeters or pocket ionization chambers designed solely for X and gamma radiation, with thick walls, or also for beta radiation, with thinner walls. Work with neutrons requires special dosimetric and control instruments based on indirect ionization. Among basic dosimetric and control instruments belong, further, dose-rate-meters, instruments for measuring and checking contamination of both the surface of the room and installations, and also especially of the hands and footwear, or the whole body. In addition, all the basic aids of personal protection must be available.

4.9.5 Radioactive Waste

After an experiment there remains radioactive waste which must be disposed of. Gaseous wastes are rather rare and we remove them by allowing them to escape through the ventilation system with an effective filter into the atmosphere. At the same time care must be taken to respect the meteorological conditions.

Radioactive liquid waste is disposed of in two ways, according to the level of activity. The simplest way in the case of a liquid with a low activity level, is to empty it into the sink. Here, however, according to regulations valid in most countries, this manner of waste disposal presumes a preceding dilution such that the activity is less than 1×10^{-4} mCi l^{-1}. Better-equipped workplaces have specially built stations to reduce the activity of the solution (by precipitation, dilution, etc.). If neither of these methods of waste disposal is practicable, the waste is collected in containers specially set apart for this purpose and transported to the site of permanent disposal.

Solid wastes are disposed of in a similar way. If they are not dissolved and diluted, they are collected in special containers and are removed by an organized agency. Short-lived radionuclide wastes are generally left to "ring down" in the laboratory.

In the majority of countries transport and disposal of radioactive waste are arranged and organized on the principles of recommendations given by the International Commission for Radioactive Protection (I.C.R.P.) and the I.A.E.A. Differences, however, may arise as regards both technological details and methods of administrative management. Hence, before starting work with radionuclides one must be acquainted with the local regulations for radioactive waste disposal.

4.9.6 Decontamination

Work with radioactive materials often necessitates decontamination of various routinely used objects. At other times some unforeseen accident or an emergency may make such decontamination urgent. Decontamination must be carried out primarily for two reasons:

A. Work in an affected area represents the danger of an internal contamination or external irradiation.

B. The use of contaminated objects during an experiment may affect the results.

Therefore, a systematic control of all objects used during work and of the working environment is indispensable, and anything found contaminated must be decontaminated. Certain objects, such as laboratory glassware, are decontaminated systematically.

Several procedures exist by which we may attain satisfactory decontamination results:

A. Mechanical removal of radioactive substances from the contaminated surfaces and decontamination by washing with tap water, pressurized water, steam, air current, suction, etc.; also scraping with fine sawdust and sand blasting.

B. Rinsing the contamined surface with solutions of acids, bases and organic solvents is a widely used procedure. The choice of the decontaminating agent will be determined by the nature of the contaminated surface and of the radioactive substance. Thus, for example, nitric acid, sulfuric acid, hydrochloric acid, potassium-, sodium-, calciumhydroxide, etc. may be used. For decontaminating nonmetal materials with a fatty surface we may employ carbon tetrachloride, chloroform, gasoline, petroleum, etc. Organic acids and their salts, too, are frequently employed for decontamination as they possess good complex-forming properties. Decontaminating agents most commonly used are citric acid and citrates, tartaric acid, acetic acid and chelates. Further substances that can be suitably employed are various detergent agents, such as ordinary soaps or special detergents. In this manner we may achieve a 99–100 % decontamination.

C. Evaporation, electrodecontamination of metals, etc., for special cases.

When working with radioactive substances as open sources, we must daily decontaminate laboratory glassware. Contaminated glassware may distort experimental results. We therefore make it a point at all times to use only decontaminated glass and, in case very low activities are involved, absolutely pure glass. The decisive factor in the selection of the decontaminating agent will be the nature of the radioactive element with which glass is contaminated. Agents most frequently employed are inorganic acids, citric acid, ammonium citrate, sodium phosphate and a whole series of other solutions. When the glassware is contaminated with short-lived radionuclides, the glass is stored and allowed to stand until its activity reduced below a certain level. When several types of radionuclides are processed in the same laboratory, separate glassware is used for each type and it is labeled with a definite color.

Special attention must be devoted to the cleaning of the hands, for in various manipulations they are most exposed to contamination with radioactive substances. Rubber gloves must always be worn when working with any radioactive material. Before taking off the gloves, we wash them thoroughly with soap under running water. If they still remain contaminated, we must clean them further according to the nature of the radioactive substance. In case of a more serious contamination it is better to discard the gloves into the waste. On removing the gloves we wash our hands thoroughly. In case of an eventual contamination of the hands an immediate cleaning must be carried out in order to prevent radioactivity from penetrating into the skin. When deciding on the manner of cleaning the hands we take into consideration the properties of the radioactive substance and the degree of contamination. Experiences so far have shown that when the contamination is not of a high degree, a thorough washing of the hands with a good quality soap and a brush in a stream of warm water is sufficient. If repeated washings fail to achieve a satisfactory decontamination of the hands, additional cleaning agents have to be employed. These differ according to the nature of the contaminant, e.g., a mixture of soap powder with sawdust, detergent soap (Marseilles soap), titanium dioxide paste, potassium permanganate solution and solutions of further agents.

4.9.7 Additional Safety Measures for Handling of Radioactive Material

All the questions relative to work with closed and open radiation sources, their storage and transport, and all the aspects concerning protection, dosimetric control, decontamination and radioactive waste disposal gathered from the experience of many countries, have been embodied in materials of the I.A.E.A. [133, 139, 145, 154, 155–160]. These materials together with recommendations by the I.C.R.P. form the basis for regulations and directives in practically every country, and these regulations are compulsory for every laboratory and every person working with radioactive material. These regulations and provisions must be accessible at every such workplace, where everyone must become familiar with them and act accordingly. In addition every workplace must have its own particular working order which supplements and interprets the general rules. This order must also include instructions for emergency cases.

References

1. ICRU: Radioactivity, Recommendations of the International Comission on Radiological Units and Measurements (1962), (ICRU) Report 10 C, Handbook 86, National Bureau of Standards, Washington 1963.
2. Tolbert B. M., Nucleonics **18** (8), 74 (1960).
3. Smith E. L., Lancet, (7069), 387 (1959).
4. Evans E. A., Stanford F. G., Nature **197** (4867), 551 (1963).
5. Evans E. A., Stanford F. G., Nature **199** (4895), 762 (1963).
6. Bourne E. J., Hutson D. H., Weigel H. J., Chem. Soc. 5753 (1960).
7. Dauben W. G., Payot P. H., J. Am. Chem. Soc. **78,** 5657 (1956).
8. Adams P. T., J. Am. Chem. Soc. **77,** 5357 (1955).
9. Rabe J. G., Guarino A., Rabe B., Int. J. Appl. Radiation Isotopes **14,** 571 (1963).
10. Apelgot S., Ekert B., Tisne M. R., *Proceedings of the Conference on Methods of Preparing and Storing Marked Molecules*, Brussels, November 13–16, 1963 (EUR 1625.e), European Atomic Energy Community, Brussels 1964, pp. 221.
11. Bayly R. J., Weigel H., Nature **188** (4748), 384 (1960).
12. Schütte H. R., *Radioactive Isotope in der organischen Chemie und Biologie*. VEB Deutscher Verlag der Wissenschaften, Berlin 1966.
13. *Proceedings of the Conference on Methods of Preparing and Storing Marked Molecules*, Brussels, November 13–16, 1963 (EUR 1625.e), European Atomic Energy Community, Brussels 1964.
14. Bendetti-Pichler A. A., Rachele J. R., Ind. Eng. Chem. Anal. Ed. **12,** 233 (1940).
15. Majer V., *Základy jaderné chemie (Principles of Nuclear Chemistry)*. SNTL, Prague 1961.
16. Haissinsky M., *La chimie nucléaire et ses applications*. Masson, Paris 1957.
17. Hensley J. W., Long A. O., Willard J. E., Ind. Eng. Chem. **41,** 1415 (1949).
18. Starik I. E., Rozovskaya N. G., ZhNKH **1,** 598 (1956).
19. Starik I. E., *Osnovy radiokhimii*. Izd. AN SSSR, Moscow 1959.
20. Lind S. C., Underwood J. E., Whittemore G. F., J. Am. Chem. Soc. **40,** 465 (1918).
21. Erbacher O., Nikitin B. A., Z. Phys. Chem. **A 158,** 216 (1932).
22. *The Radiochemical Manual* (ed. 2). The Radiochemical Centre, Amersham 1966.
23. Broda E., Schönfeld T., *Radiochemische Methoden der Mikrochemie*. Springer-Verlag, Vienna 1955.
24. Hahn O., *Applied Radiochemistry*. Cornell Univ. Press, New York 1936.
25. Booth A. H., Trans. Faraday Soc. **47,** 633, 640 (1951).
26. Hahn O., Z. Phys. Chem. **A 144,** 161 (1929).
27. Nefedov V. D., *Adsorptsiya radioaktivnykh elementov na ionnykh krystalakh*. In Ref. 28.
28. Murin A. N., Nefedov V. D., Shvedov V. D., (Ed.), *Radiokhimiya i khimiya yadernykh protsessov*. Gos. nauchno-tekhn. izd. khim. lit., Leningrad 1966.
29. Wahl A. C., Bonner N. A., *Radioactivity Applied to Chemistry*. Wiley, New York 1951.
30. Khlopin V. G., Merkulova M. C., DAN SSSR **65,** 861 (1949).
31. Melikhov I. V., *Izuchenie protsessov perekhoda mikroelementov iz zhidkoy fazy v tverduyu*. In *Sossazhdenie i adsorptsiya radioaktivnykh elementov*. Nauka, Moscow 1965.
32. Paneth F., Z. Phys. Chem. **89,** 513 (1915).
33. Paneth F., Ber. **43,** 1215 (1924).
34. Fajans K., Beckerath H., Z. Phys. Chem. **97,** 478 (1921).
35. Fajans K., Erdey-Gruz T., Z. Phys. Chem. **A 158,** 97 (1932).
36. Hahn O., Ber. **59,** 2014 (1926).

37. Hahn O., Naturwiss. **14,** 1197 (1926).
38. Trudy RIAN **1** (1930), **2** (1933), **4** (1938).
39. Khlopin V. G. et al., Z. anorg. Chem. **143,** 97 (1925); **166,** 311 (1927); **172,** 310 (1929).
40. Khlopin V. G., *Izbrannye trudy.* T. 1, Izd. AN SSSR, Moscow 1957.
41. Doerner H. A., Hoskins W. M., J. Am. Chem. Soc. **47,** 662 (1925).
42. Haissinsky M., *Les radiocolloïdes.* Actualités scient. et ind. No. 185, Hermann, Paris 1935.
43. Starik I. E., Z. Phys. Chem. **A 157,** 269 (1931).
44. Werner O., Z. Phys. Chem. **A 156,** 89 (1931).
45. Melish C. E., Payne J. A., Worral G., Radiochim. Acta **2,** 204 (1964).
46. Nefedov V. D., *Kolloidnoe sostoyanie radioaktivnykh elementov.* In Ref. 28.
47. Grahame D. C., Seaborg G. T., J. Am. Chem. Soc. **60,** 2524 (1938).
48. McCallum K. J., Hockovsky S. A., J. Chem. Phys. **16,** 255 (1948).
49. Studier M. H., Hyde E. K., Phys. Rev. **74,** 591 (1948).
50. Overman R. T., Clark H. M., *Radioisotope Techniques.* McGraw-Hill, New York 1960.
51. Chase G. D., Rabinowitz J. L., *Principles of Radioisotope Methodology* (ed. 3). Burgess 1964.
52. Bethe J. A., Revs. Modern Phys. **22,** 213 (1950).
53. Brown R. E., Jarmie N., *Index and Annotated Bibliography of Range and Stopping Cross Section Data.* Los Alamos National Laboratory Report LA-2156.
54. Whaling W., *The Energy Loss of Charged Particles in Matter.* In S. Flügge, (Ed)., *Handbuch der Physik.* Vol. XXXIV, Springer-Verlag, Berlin 1958.
55. Siegbahn K. (Ed.), *Alpha, Beta and Gamma-Ray Spectroscopy.* North Holland, Amsterdam 1964.
56. Libby W. F., Phys. Rev. **69,** 671 (1946).
57. Libby W. F., Anal. Chem. **19,** 2 (1947).
58. Glendenin L. E., Nucleonics **2,** 12 (1948).
59. Feather N., Proc. Cambridge Phil. Soc. **34,** 599 (1938).
60. Sargent B. W., J. Res. **17,** 82 (1939).
61. Katz L., Penfold A. S., Rev. Mod. Phys. **24,** 28 (1952).
62. Flammersfeld A., Z. Naturforsch. **2a,** 370 (1947).
63. Bleuler E., Zünti W., Helv. Phys. Acta **19,** 137, 375 (1946).
64. Kuczera J., Nukleonika **12,** 347 (1967).
65. Evans R. D., *The Atomic Nucleus.* McGraw-Hill, New York 1955.
66. Tittle C. W., Tech. Bull. No. 8, Nuclear-Chicago Corp., Des Plaines, Ill. 1960.
67. Bleuler E., Goldsmith G. J., *Experimental Nucleonics.* Rinehart, New York 1952.
68. Henderson L., Kracek F., J. Am. Chem. Soc. **49,** 738 (1927).
69. Spicyn V. I. et al., *Metody raboty s primeneniyem radioaktivnykh indikatorov.* Izd. AN SSSR, Moscow 1955.
70. Harley J. H., Hallden N., Nucleonics **13,** (1), 32 (1955).
71. Hayashi S., Otozai K., Isotopes and Radiation **3,** 83 (1960).
72. Otozai K., Hayashi S., Int. J. Appl. Radiation Isotopes **16,** 681 (1965).
73. Watt J. S., Int. J. Appl. Radiation Isotopes **15,** 617 (1964).
74. Clayton C. G., Cameron J. F., *A Review of the Design and Application of Radioisotope Instruments in Industry.* In *Radioisotope Instruments in Industry and Geophysics* (Vol. 1). IAEA, Vienna 1966, p. 15.
75. Lambie D. A., *Techniques for the Use of Radioisotopes in Analysis (A Laboratory Manual).* Spon, Ltd., London 1964.
76. Martell A. E., Calvin M., *Chemie kovových chelátů* (Transl. from English). Nakl. ČSAV, Prague 1959.

77. Starý J., *The Solvent Extraction of Metal Chelates*. Pergamon Press, London 1964.
78. Morrison G. H., Freiser H., Anal. Chem. **30,** 362 (1958); **32,** 37 R (1960); **34,** 64 R (1962); **36,** 93 R (1964).
79. Morrison G. H., Freiser H., *Solvent Extraction in Analytical Chemistry*. Wiley, New York 1957.
80. Mizuike A., *Separations and Preconcentrations*. In G. H. Morrison (Ed.), *Trace Analysis—Physical Methods*. Wiley, New York 1965.
81. Furman N. H., Mundy R. J., Morrison G. H., U.S. A.E.C. Report AECD-2861.
82. Joliot F., J. Chim. Phys. **27,** 119 (1930).
83. Lingane J. J., *Electroanalytical Chemistry* (ed. 2). Interscience, New York 1958.
84. Tanaka N., Kolthoff I. M., Elving P. J., *Treatise on Analytical Chemistry* (Part 1, Vol. 4). Interscience, New York 1963.
85. Tekster E. N., *Elektrokhimiya radioaktivnykh elementov*. In Ref. 28.
86. Haissinsky M., *Electrochimie des substances et des solutions extrêmement diluées*. Hermann, Paris 1946.
87. Rogers L. B., Anal. Chem. **22,** 1386 (1950).
88. Radin N. S., Tech. Bull. No. 7, Nuclear-Chicago Corp., Des Plaines, Ill. 1960.
89. Campbell G., Glastonbury A. H., Stevenson M. D., Nature **182,** 1100 (1958).
90. Isbell H. S., Frush H. L., Peterson R. A., J. Res. Nat. Bur. Stds, **63 A**, 171 (1959).
91. Jervis R. E., Talanta **2,** 89 (1959).
92. Bronner F., Jernberg N. A., Anal. Chem. **29,** 462 (1957).
93. Burr W. W., Marcia J. A., Anal. Chem. **27,** 571 (1955).
94. Oppermann R. A., Nystrom R. F., Nelson W. O., Brown R. E., Int. J. Appl. Radiation Isotopes **7,** 38 (1959).
95. Radin N. S., Fried R., Anal. Chem. **30,** 1926 (1958).
96. Herberg R. J., Anal. Chem. **32,** 42 (1960).
97. Brown W. O., Badman H. G., Biochem. J. **78,** 571 (1961).
98. Radin N. S., Tech. Bull. No. 11, Nuclear-Chicago Corp., Des Plaines, Ill. 1961.
99. Gordon C. F., Wolfe A. L., Anal. Chem. **32,** 574 (1960).
100. Heyssel R. M., J. Clin. Invest. **40,** 2134 (1961).
101. Kerr V. N., Hayes F. N., Ott D. G., Int. J. Appl. Radiation Isotopes **1,** 284 (1957).
102. Birks J. B., *Theory and Practice of Scintillation Counting*. Pergamon Press, London 1964.
103. Schram E., *Organic Scintillation Detectors*. Elsevier Publishing Co., Amsterdam 1963.
104. Eidinoff M. L., Knoll J. E., Science **112,** 250 (1950).
105. Jackson F. L., Lampe H. W., Anal. Chem. **28,** 1735 (1956).
106. Freedman A. J., Anderson E. C., Nucleonics **10** (8), 57 (1952).
107. Bradley J. E. S., Holloway R. C., McFarlane A. S., Biochem. J. **57,** 192 (1954).
108. Ganourel J. D., Baker M. J., Koch C. W., Anal. Chem. **27,** 795 (1955).
109. Körbl J., Mikrochim. Acta 1705 (1956).
110. Večeřa M., Synek L., Collect. Czech. Chem. Commun. **23,** 1202 (1958).
111. Anderson R. C., Delabarre Y., Bothner-By A. A., Anal. Chem. **24,** 1298 (1952).
112. Glascock R. F., *Isotopic Gas Analysis for Biochemists*. Academic Press, New York 1954.
113. Chistman D. R., Stuber J. E., Bothner-By A. A., Anal. Chem. **28,** 1345 (1956).
114. Gordon A. S., Heimel S., J. Am. Chem. Soc. **73,** 2942 (1951).
115. Van Slyke D. D., Folch J., J. Biol. Chem. **136,** 509 (1940).
116. Peters J. H., Gutman H. R., Anal. Chem. **25,** 987 (1953).
117. Moyer J. D., Isbell H. S., Anal. Chem. **29,** 393 (1957).
118. Katz J., Abraham S., Baker N., Anal. Chem. **26,** 1503 (1954).
119. Banks T. E., Crawhall J. C., Smyth D. G., Biochem. J. **64,** 411 (1956).

120. Biggs M. W., Kritchewski D., Kirk M. R., Anal. Chem. **24,** 223 (1952).
121. Wilzbach K. E., Kaplan L., Brown W. G., Science **118**, 522 (1953).
122. Cameron J. F., Nature **176,** 1264 (1955).
123. Drever R. W. P., Moljk A., Rev. Sci. Instr. **27,** 650 (1956).
124. Eidinoff M. L., J. Am. Chem. Soc. **69,** 2504 (1947).
125. Swain C. G., Kreiter V. P., Sheppard W. A., Anal. Chem. **27,** 1157 (1955).
126. Henly J. W., Schwandiman L. C., Radiation Research **4,** 278 (1956).
127. Graff J., Rittenberg D., Anal. Chem. **24,** 878 (1952).
128. Robinson C. V., Rev. Sci. Inst. **22,** 353 (1951).
129. Wing J., Joston W. H., Science **121**, 674 (1955).
130. Mai K. L., Babb A. U., Nucleonics **13**, (2), 52 (1955).
131. Balis M. E. et al., J. Biol. Chem. **188,** 217 (1951).
132. Rajewsky B., *Strahlendosis und Strahlenwirkung.* Thieme, Stuttgart 1956.
133. International Atomic Energy Agency: *Safe Handling of Radioisotopes.* Medical Addendum (IAEA Safety Series No. 3), IAEA, Vienna 1960.
134. International Commission on Radiological Protection: *Radiation Protection, Recommendations of the International Commission on Radiological Protection, Report of Committee II on Permissible Dose for Internal Radiation (1959).* ICRP Publication 2, Pergamon Press, Oxford 1960.
135. International Commission on Radiological Protection: *Radation Protection, Recommendations of the International Commission on Radiological Protection, Report of Committee III on Protection against X-Rays up to Energies of 3 MeV and Beta- and Gamma-Rays from Sealed Sources (1960).* ICRP Publication 3, Pergamon Press, Oxford 1960.
136. International Commission on Radiological Protection: *Radiation Protection, Recommendations of the International Commission on Radiological Protection, Report of Committee IV (1953–1959) on Protection against Electromagnetic Radiation above 3 MeV and Electrons, Neutrons and Protons* (adapted 1962, with revisions adapted in 1963). ICPR Publication 4, Pergamon Press, Oxford 1964.
137. International Commission on Radiological Protection: *Radiation Protection, Recommendations of the International Commission on Radiological Protection, Report of Committee V on the Handling and Disposal of Radioactive Materials in Hospitals and Medical Research Establishments (1964),* ICRP Publication 5, Pergamon Press, Oxford 1965.
138. International Commission on Radiological Protection: *Radiation Protection, Recommendations of the International Commission on Radiological Protection* (as amended 1959 and revised 1962). ICRP Publication 6, Pergamon Press, Oxford 1964.
139. International Atomic Energy Agency: *Safe Handling of Radioisotopes.* First Edition with Revised Appendix 1 (IAEA Safety Series No. 1), IAEA, Vienna 1962.
140. Health Physics **3,** 1 (1960).
141. Price B. T., Horton C. C., Spinney K. T., *Radiation Shielding.* Pergamon Press, London 1957.
142. Goldstein H., Wilkins J. E., Jr., *Calculations of the Penetration of Gamma Rays.* NYO—3075, U.S. A.E.C.
143. Tittle C. V., Tech. Bull. No. 9., Nuclear-Chicago Corp., Des Plaines, Ill. 1961.
144. Gusev N. D., *Spravochnik po radioaktivnym izlucheniyam i zashchite.* Medgiz, Moscow 1956.
145. International Atomic Energy Agency: *Safe Handling of Radioisotopes. Health Physics Addendum* (IAEA Safety Series No. 2), IAEA, Vienna 1960.
146. Balderston J. L., Taylor J. J., Brucker G. J., *Nomograms for the Calculation of Gamma Shielding.* AECD No. 2934, 1950.
147. Jeager T., *Technischer Strahlenschutz.* Thieme, Munich 1959.

148. Sherwood R. J., *A Short Course in Radiological Protection.* (AEREL/101), Harwell 1959.
149. Lindachers K. H., *Praktische Durchführung von Abschirmungsberechnungen.* Thieme, Munich 1961.
150. West R., Report RCC-R 165, Radiochemical Centre, Amersham 1963.
151. Kimel' L. R., Mashkovich V. P., *Zashchita ot ioniziruyushchikh izlucheniy* (Spravochnik). Atomizdat, Moscow 1966.
152. Rockwell T., *Reactor Shielding Design Manual* (ed. 3). McGraw-Hill, New York 1956.
153. Goldstein H., *Fundamental Aspects of Reactor Shielding.* Addison-Wesley, Reading, Mass. 1959.
154. International Atomic Energy Agency: *A Basic Toxicity Classification of Radionuclides* (IAEA Technical Reports Series No. 15), IAEA, Vienna 1963.
155. International Atomic Energy Agency: *Regulation for the Safe Transport of Radioactive Materials* (IAEA Safety Series No. 6), IAEA, Vienna 1961.
156. International Atomic Energy Agency: *The Use of Film Badges for Personnel Monitoring* (IAEA Safety Series No. 8), IAEA Vienna 1962.
157. International Atomic Energy Agency: *Basic Safety Standards for Radiation Protection* (IAEA Safety Series No. 9), IAEA, Vienna 1962.
158. International Atomic Energy Agency: *The Management of Radioactive Wastes Produced by Radioisotope Users* (IAEA Safety Series No. 12), IAEA, Vienna 1965.
159. International Atomic Energy Agency: *The Provision of Radiological Protection* (IAEA Safety Series No. 13), IAEA, Vienna 1965.
160. International Atomic Energy Agency: *The Basic Requirements for Personal Monitoring* (IAEA Safety Series No. 14), IAEA Vienna 1965.

5. Principles of Nuclear Tracer Methods

Š. VARGA

5.1 Introduction

A tracer or indicator method, used in research, consists of labeling elements or systems forming the object of further studies. The basic idea of a tracer method in chemistry dates back a long time [1]. Among its initial forms of implementation belongs the use of dyes which are added in minute quantities to the system under study. The choice of the dyes is determined by certain criteria, the basic condition being that the added dye change its color sharply at a certain well-defined stage of the chemical reaction. The presence of the dye, however, must not interfere with the physical and chemical properties of the system. This condition is fulfilled, as a rule, for the amount of the dye added to the system is usually negligible and its lower limit depends solely on a reliable identification of its color.

Isotope tracers, both by stable and radioactive, meet the same two criteria. Usually they are applied in considerably lower concentrations than the dyes, since the limit of their detectability is extremely low.

They can be suitably applied to indicate elements and compounds, or atoms and molecules. When employed directly as an indicator, a radioisotope may be considered as an atom tracer and serves to label atoms of the isotope element.

When a radionuclide becomes incorporated into a molecule that contains its isotopic element, a radioactive labeled molecule is obtained—a molecule tracer—that serves to indicate molecules of an isotope compound.

The admixtures used in indicator methods in the function of tracers may have properties very similar to or very different from the basic substance. In the case of dye indicators generally only one or some few common properties are involved and hence dyes represent only imperfect tracers. Thus, for instance, in the case of phase transformations (evaporation, crystallization, etc.) the staining agent does not follow the substance.

An ideal tracer, which cannot be achieved in practice, would have an identity of all its properties and those of the basic substance. However, in such a case there would be no possibility of an individual identification of both the substances. Both the substances would be alike.

The nearest to the perfect tracer element are nuclear isomer tracers. Trace amounts of radioactive indicators do not alter the thermodynamic and kinetic properties of the labeled substance in the system and the emittance of high energy radiation enables their detection even in extremely minute quantities. Another priority of isotopic indicators resides in the fact that by their means we can directly label a particular element, regardless of its bond type within the molecule.

5.2 Historical Review of Radioindicator Isotope Methods

The origins of tracer methods in isotope studies date back to the early years of the present century, and are related to Rutherford's laboratories in Manchester, where Hevesy worked to separate radium D from an enormous amount of lead. Experiments showed that a chemical separation of these substances was unsuccessful and these elements were later designated as isotopes. However, the failure of these efforts led Hevesy to the following idea: If RaD cannot be separated from Pb, an admixture of RaD can be used to "label" the originally inactive lead.

This significant idea was further developed in the Radium Institute of Vienna by Hevesy and Paneth. Since isotopes cannot be separated by ordinary chemical methods, they labeled lead with its radioactive isotope (RaD), the value of the specific radioactivity being known. They were then able to detect even minute amounts of lead, through its radioactive radiation, and also could perform a quantitative measurement of the element.

The first studies by Hevesy and Paneth were not really analytical in the strict sense of the word, why they were concerned with the estimation of the solubility of lead chromate and lead sulfide. The method developed by them traced out the way for analytical methods based on radioactivity measurement and in particular for the development of analytical procedures involving the use of radioactive labeled substances [2, 3]. The introductory lines of this historical paper are as follows [3]: "The fourth degradation product of radium emanation RaD shows, as is known, all the chemical reactions of lead: if RaD is mixed with lead, the latter can no longer by separated by any physical or chemical method. Once both the substances become thoroughly mixed, then their concentration ratio is the same even in the case of the least amounts of lead separated from the solution. As a result of the activity of RaD its incomparably smaller quantities than those of lead can be determined, and when added to lead it can serve for its qualitative proof and quantitative determination, and thus RaD behaves as an 'indicator' of lead."

5.3 Principles of Radioactive Indication of Chemical Systems

The physicochemical properties of isotopes are not essentially affected by their stable or radioactive nature. This means of course that the behavior of stable nuclides may be inferred from that of the radioactive isotopes of the same element. Hence, radioactive isotopes may be utilized to follow the behavior of an element or its compound in a chemical system in the course of certain chemical processes [1, 4–9, 29]. A characteristic feature of labeling by means of radioisotopes is that the rate of radioactive disintegration is not affected by the bond type of the corresponding atoms. This means that radiation detection provides a substantially more reliable test of the presence of an element than the majority of the other physical and chemical procedures. The presence of a radioactive indicator can be proved almost regardless of the form of its chemical bond. A radioactive isotope of the relevant element, which is in reality a mixture of its stable isotopes, is added to the system. In this way, the element under study becomes radioactively labeled. Hence, radioactive isotopes of a studied element are radioactive indicators, tracers in the particular system. A radioactive indicator is generally added to the system in the same chemical form as that of the element to be labeled.

However, when a radioactive indicator is utilized, care must be taken for its chemical and radiochemical purity. That means that radioactive indication refers to only one type of molecule, or that the radioactive indicator is present in the labeled system in only one bond type.

The use of radioactive indicators can be schematically illustrated as follows [10]. Let the chemical system (S) which is to be indicated contain, besides others, the element (A) to be followed, S(A, ...), and let the system contain x atoms with stable isotopes of the element A, these atoms being generally present in a macroscopic amount. To the system we add y atoms with radioactive isotope *A of the element studied: the system can therefore be characterized as

$$S(xA, y{*}A, \ldots),$$

where, however, radioactive atoms are present in the form of a macrocomponent.

The basic principle in the application of radioactive indicators is the assumption that every change observed on radioactive atoms *A applies also to stable atoms A. Let us assume that in a chemical reaction y' radioactive atoms *A pass from the initial system S into the reaction product P. This means also that an identical ratio of stable atoms A passes from the initial system into the reaction product, hence,

$$x' = x\frac{y'}{y}. \tag{5.1}$$

Schematically:

$$\mathrm{S}(x\mathrm{A},\, y{*}\mathrm{A},\, \ldots) \rightarrow \mathrm{P}(x'\mathrm{A},\, y'{*}\mathrm{A},\, \ldots). \tag{5.2}$$

A similar consideration would enable us to determine the ratio of inactive atoms A which pass from the initial system into the reacting intermediate product following the reaction time t.

The use of radioactive indicators is based on the assumption that the inactive and the radioactive isotopes are homogeneously distributed in the indicated chemical system and that this distribution does not change during the course of the processes. Under these conditions the presence of radioactivity in a portion of the system indicates the presence of the element studied. Detection of radioactivity constitutes in this case a qualitative proof of the labeled element, or of its compound. The value of radioactivity, on the other hand, indicates the quantity of the element or compound studied, in the given portion of the labeled system. Therefore, this value serves for the quantitative determination of the element under study.

5.3.1 Physical and Chemical Properties of Isotopes

The statement that the physical and chemical properties of isotopes are identical is but an approximation. In reality isotopes and even their compounds behave differently as regards their physical and chemical properties. Molecules in which atoms of one or several elements, with the same arrangement and identical structure, are isotopes to one another are called isotope molecules. The general indication of the differences in the properties of isotopes and their compounds is the *isotope effect*.

Isotope effects are indeed extremely small, even negligible, yet they are adequate for the separation of isotopes. These effects are conditioned by a difference in isotope mass, a special nucleon arrangement in them, their different spin and a whole series of further specific nuclear properties.

The isotope effect increases as the mass ratio of the corresponding isotopes deviated from unity. It greatest significance is in the case of the lightest elements and their compounds. Relative differences in isotope mass of certain elements are given in Table 5.1.

The isotope effect is more pronounced if elements are concerned since in their compounds further atoms in the molecule reduce the effect. The heavier the atoms, the more the reduction.

The significance of the isotope effect decreases with increasing atomic number. A measurement of the isotope effects requires the use of sensitive experimental methods even for light elements; even the most sensitive ones are inadequate when heavy elements are involved.

Table 5.1

Relative Mass Differences of Some Nuclides

Nuclide	Mass related to the most significant isotope (%)
H (^{1}H)	—
D (^{2}H)	+100
T (^{3}H)	+200
^{12}C	—
^{13}C	+ 8.3
^{14}C	+16.7
^{31}P	—
^{32}P	+3.2
^{33}P	+6.4
^{234}U	-1.7
^{235}U	-1.3
^{238}U	—

Table 5.2

Density Ratio of Hydrogen Isotopes and Compounds

A	B	Temperature (°C)	Density ratio d_A/d_B
D_2	H_2	-254	2.242
T_2	H_2	-253	3.560
D_2O	H_2O	25	1.10775
D_2O_2	H_2O_2	25	1.058
ND_3	NH_3	25	1.349

Isotopes differ in the number of neutrons; consequently the basic physical difference between them resides in their specific weight. This difference is so striking in the case of hydrogen isotopes that deuterium is also known as heavy hydrogen and its oxide as heavy water.

Table 5.2 [12] shows a comparison of the density ratio of hydrogen and its compounds.

According to its thermodynamic properties isotopic composition may manifest itself by different values of the pressure of saturated vapors, temperature and heat of phase transformations, thermal capacity in the temperature dependence of thermodynamic parameters, etc.

Different values of the chemical potential of isotope molecules may become manifest in the value of the equilibrium constant. The effect of ^{14}N and ^{15}N isotopes manifests itself, for instance, in the system

$$^{15}NH_3 + ^{14}NH_4^+ \rightleftharpoons ^{14}NH_3 + ^{15}NH_4^+ \tag{5.3}$$

in the value of the equilibrium constant:

$$K = \frac{[^{14}NH_3][^{15}NH_4^+]}{[^{15}NH_3][^{14}NH_4^+]} \tag{5.4}$$

which, in consequence of the different values of the chemical potentials in the isotope compounds, deviates from unity. At 25 °C, $K = 1.035$ [11]. Isotopes and their compounds can differ also in viscosity coefficient and in other physical constants.

There are also differences in solubilities, if we compare the data obtained for some inorganic compounds in water and in heavy water. See Table 5.3 [12].

Similarly, differences can be observed in the solubility of organic compounds; or in ion mobilities in isotope electrolytes. The isotope composition may cause a change in the degree of electrolyte dissociation, in the electrode potential, etc.

The differences in physicochemical properties shown in Table 5.2 were observed with hydrogen isotopes and their compounds. Isotope effects in the other elements are less pronounced, but still measurable, for instance, in the case of helium, lithium and carbon isotopes.

Differences in the mass of isotope nuclei result in changes in both atomic and molecular spectra. Different nuclear spins of isotopes bring about changes in the fine nuclear structure. The isotopic composition of hydrogen also affects the hydrogen bonds. Among the optical properties, optical activity and molar refraction can be affected by isotope composition.

The isotopic action may become manifest also by stereoisomerism of the isotope molecule [11].

Particular attention must be paid to the reaction kinetic action of an isotope. Kinetic isotope effects are mainly involved in the rate of reaction, conditioned principally by the value of the activation energy and the frequency factor. The different reaction rates of isotope compounds are reflected also in the value of the equilibrium constant [13, 14]. Kinetic and thermodynamic isotope effects diminish with increasing temperature.

The following consideration will provide a closer view of the kinetic isotope effects [12, 15–18]. Given: a reaction within the framework of which a mixture of isotope molecules AB and *AB disintegrates or reacts with further molecules C, D, ... thus giving rise to products P and *P, according the schema

$$AB + C + \ldots \rightarrow P, \tag{5.5}$$

$$^*AB + C + \ldots \rightarrow {}^*P. \tag{5.6}$$

Table 5.3

Solubility Data of Some Salts in H_2O and D_2O

Compound	Temperature (°C)	Solubility [Mol per Mol of solvent] L_{H_2O}	L_{D_2O}	Δ	$\frac{\Delta}{L_{H_2O}}$ 100
KCl	0	6.81	5.69	1.12	16.4
	25	8.65	7.80	0.76	8.8
	100	13.54	13.06	0.48	3.5
	180	18.44	18.12	0.32	1.7
CdI_2	25	4.34	3.21	1.13	26
NaBr	5	3.25	2.87	0.38	11.9
$K[Ag(CN)_2]$	5	1.290	0.974	0.316	24.5
$K_2Cr_2O_7$	5	0.358	0.236	0.122	33.5
	25	0.955	0.712	0.243	26.6
	35	1.372	1.067	0.305	23.0
$KClO_3$	5	0.585	0.512	0.073	12.4
$Hg(CN)_2$	5	0.537	0.424	0.113	21
$PbCl_2$	25	0.071	0.045	0.026	36

The rate laws for the above reactions are the following:

$$-\frac{d[AB]}{dt} = k[AB]\,[C]\ldots, \tag{5.7}$$

$$-\frac{d[{}^*AB]}{dt} = {}^*k[{}^*AB]\,[C]\ldots, \tag{5.8}$$

and hence

$$\frac{d[AB]}{d[{}^*AB]} = \frac{k}{{}^*k}\,\frac{[AB]}{[{}^*AB]}, \tag{5.9}$$

or

$$d\ln[AB] = \alpha\, d\ln[{}^*AB], \tag{5.10}$$

where $\alpha = k/{}^*k$ is the ratio of the rate constants and is a measure of the kinetic isotope effect. If $\alpha = 1$, then the rate of reaction is not affected by the isotope composition of the starting reaction components.

Let the total concentration of the isotope compound AB + *AB in the atomic fractions at the start of the reaction ($t = 0$) be a_0, and at time t be a; hence the content of isotope *A x_0 and x.

Integrating Eq. (5.10) between $t = 0$ and t we obtain

$$\ln \frac{a(1-x)}{a_0(1-x_0)} = \alpha \ln \frac{ax}{a_0 x_0}, \tag{5.11}$$

where

$$[\mathrm{AB}] = a(1-x), \tag{5.12}$$

$$[{}^*\mathrm{AB}] = ax. \tag{5.13}$$

On introducing the fraction $s = \dfrac{a_0 - a}{a_0}$ into Eq. (5.11) and after rearranging we obtain

$$-\frac{\alpha - 1}{\alpha} \ln (1-s) + \frac{1}{\alpha} \ln \frac{1-x}{1-x_0} = \ln \frac{x}{x_0}. \tag{5.14}$$

If a radioactive indicator is used (its amount is usually very small), $x_0 \ll 1$; then Eq. (5.14) can be simplified into

$$-\frac{\alpha - 1}{\alpha} = \frac{\ln \dfrac{x}{x_0}}{\ln (1-s)}. \tag{5.15}$$

Equations (5.14) and (5.15) permit the value of the kinetic isotope effect (α) to be calculated from changes in the isotope composition (from x_0 till x) even when the reaction is incomplete.

The magnitude of the kinetic isotope effect can also be calculated from the isotope composition (y) of the reaction product.

The relation for the material balance of isotope *A in the reaction product and the unreacted rest

$$y(a_0 - a) + xa = x_0 a_0 \tag{5.16}$$

is as follows. After introduction of fraction s, we got

$$y = \frac{x_0 - x(1-a)}{s}. \tag{5.17}$$

By means of this relation, x is removed from Eq. (5.14) and in the case of the comparable amounts of isotopes A and *A, we obtain:

$$\frac{1}{\alpha} \ln \frac{1 - x_0 - s(1-y)}{1 - x_0} = \ln \left(1 - \frac{ys}{x_0}\right). \tag{5.18}$$

Similarly, for negligible amounts of isotope *A from relation (5.15) we obtain

$$\frac{1}{\alpha} = \frac{\ln\left(1 - \dfrac{ys}{x_0}\right)}{\ln(1 - s)}. \tag{5.19}$$

These expressions may be considerably simplified if the isotope composition of the reaction product is determined in its quotient

$$\alpha = \frac{x_0}{y} \frac{1 - y}{1 - x_0}, \tag{5.20}$$

or in the case of trace amounts of radioactive indicators

$$\alpha = \frac{x_0}{y}. \tag{5.21}$$

The significance of the isotope effect for an analytical application of radioactive indicators depends on the applied radioactive isotope. The isotope effect depends on the mass difference of the isotopes and becomes manifest in the change of the reaction rate of the isotopes, for their binding energy in the molecule decreases with decreasing mass. The carbon-hydrogen bond can be far more easily ruptured than the carbon-tritium bond [30].

5.3.2 Limitations of Radioactive Indication

Along with its evident advantages, the use of isotopes and in particular of radioactive isotopes has certain limitations [10–12, 19–21].

A primary limitation is due to isotope effects. Because of different mass and a different arrangement of nucleons in isotopic atom nuclei, the physical and chemical properties of isotope atoms and isotope molecules are not identical. When light nuclides are used as indicators, it is not always possible to disregard the kinetic isotope effects. Isotope effects are dealt with in more detail in Chapt. 5.3.1.

A further limitation in the use of radioactive indicators is represented by their radiation action on the labeled system.

Theoretically, chemical transformations due to the action of radiation take place in every radioactively indicated system. As a matter of fact, in every such system there occurs an interaction between radiation emitted by the indicator isotope and the atoms and molecules of the labeled system. The radiation energy is transferred to the system either through ionization or excitation. As result of the interaction highly reactive chemical species (e.g., free radicals, excited molecules) are formed which through further processes may yield a number of undesirable products.

In aqueous solutions also redox reactions may proceed. Organic compounds may decompose (radiolysis) or undergo polymerization or isomerization.

One of the possible sources of errors from the viewpoint of purity of the radioactive indicator is its autoradiolysis. This phenomenon may be encountered due to high specific activity or long storage, in particular when organic molecules are used as radioactive tracers.

The action of radiation may generally be neglected if the absolute activity of the radioactive indicator or the specific activity of the labeled system is low. On the basis of known values of the radiation chemical yield in the majority of chemical systems, the activity of radioactive indicators applied in analytical studies is, as a rule, of such an order that the radiation effect is unobservable.

These phenomena, however, cannot be neglected in certain cases in biological systems whose radiation sensitivity may be extremely high. Radiation sensitivity is relatively high also in the colloid systems.

The effect of radiation on the labeled chemical system is experimentally verified by a change in its specific activity after a sufficiently long interval. At the same time the decrease in specific activity due to the radioactive decay of the indicator cannot be neglected. If the results of the tests remain unchanged, the effect of radiation on the indicated system may be neglected.

Further disturbing effects in the use of radioactive indicators concern chemical transformations due to radiation. In a molecule labeled with a radioactive element of one of its elements, there ensues during nuclear transformation a nuclide of a new element, and hence a new molecule is formed, a molecule of a different compound. Besides this phenomenon, at the instant an atomic nucleus disintegrates, it gives rise to a recoiled nuclide which is capable of acquiring sufficient energy to induce ionizing and excitating effects like those of radioactive radiation.

Radioisotope tracer atoms are detected on the basis of their radioactivity. After their decomposition they cannot be detected any more and their subsequent fate in the system is of no further interest. If the daughter atoms themselves are radioactive we have to consider this fact when choosing the appropriate detection method. Such a multistep radioactive transformation can be met with in members of radioactive series. Their members are related genetically and we recognize mother and daughter radioactive nuclides. These problems should be considered when natural radioactive indicators are used. For example, if RaE(^{210}Bi) is used as a tracer for bismuth, alpha particles from the daughter ^{210}Po should be eliminated. In the case of tracer UX_1, which is a tracer for thorium, the radiation of its daughter, UX_2, is detected: The reason is that the short half-life of UX_2 (1.18 min) assures that it is in a transient equilibrium with UX_1 at the time when the sample activity is measured, and hence the total activity is proportional to the UX_1 content. Numerous multiple disintegrations occur in the case of fission products. If, for instance, ^{140}Ba – half-life 12.8 days – is used as indicator for barium, isolated samples should be separated chemically from the daughter product ^{140}La – half-life 40 hr – or stored for a week or two before measurement, in order that a transient radioactive equilibrium be attained. Activities of isomer transitions represent interesting cases; if the isomer ^{80m}Br – half-life 270 min – is chosen as the indicator for bromine, the lower isomer state ^{80}Br – half-life 18 min – will be permanently present, and since the isomer transition is accompanied by chemical effects, these two isomers can always be present in various chemical forms. Hence the four-and-a-half hour ^{80m}Br can be used with certainty, it being only necessary that the isolated samples be measured after a time interval longer than the half-life of ^{80}Br. These special cases do not arise with the majority of currently used isotope indicators [1].

Applicability of radioactive indicators may be limited by the exchange reaction between the radioactive isotope in the labeled compound and isotopes of the indicated element in the non-labeled one. This refers principally to organic sulfur compounds.

The isotope indicator when added to a particular system is diluted by the non-radioactive isotopes of the same element. Various analytical procedures (precipitation, extraction, titration, etc.) further dilute the isotopic indicator. This multiple dilution reduces the efficiency of the detection.

With a 1 % content of the stable isotope indicator in the labeled system – in view of the detection sensitivity of stable isotopes – the dilution of the indicator in analytical procedures cannot be higher than 10^3–10^4 fold. Radioactive indication under the same conditions admits a dilution of the indicator as much as 10^6 times higher [22].

A particular case of limitation of isotope indication occurs when there is question of a microindication of a component which is itself present in the studied system as a microcomponent, or even in only a trace amount. The addition of the indicator even though in a negligible trace amount can affect the processes that relate to the labeled substance, for it is likewise present as a microcomponent.

5.3.3 Technique of Radioactive Indication

Labeling by means of radioactive isotopes may be carried out in various technical procedures according to the size of the indicated system [23–27]. The procedures are designated as macroindication and microindication, according to whether macro- or microscopic systems are involved. Macroindication refers to the labeling of the various objects or their parts, but in the biological systems also of individuals or their organs. Microindication is also called atomic, or molecular indication, as for radioactive indication refers to atomic or molecular systems hence, to elements or compounds.

Macro- and microindication make use predominantly of the technique of external indication in which the radioactive indicator is added in a suitable chemical and physical form to the labeled system.

Macroindication generally makes use of a local labeling of the object studied, which fully satisfies the needs of following the mechanical motion of the labeled object; at the same time no homogeneous dispersion of the indicator takes place within the labeled system (a study of the wear of machine parts, of the fire-resistant tiling in glass-, steel-, or blast-furnaces). In the case of macroscopic indication of living organisms, we generally presume a homogeneous distribution of radioactively labeled individuals among non-labeled ones (a study of insect movement or other living organisms).

Microindication, i.e., labeling of atoms and molecules requires, on the contrary, a homogeneous reciprocal distribution of both radioactive and stable isotopes in the studied element of the labeled system.

Atomic and molecular indication may be classified into several types, each with a particular procedure of the external technique of labeling.

The ideal requirements are closely fulfilled by the nuclear isomeric (isobar–isotope) indication. As is known, radioactive indicator atoms and labeled atoms have the same mass, and therefore the isotope effects cannot be observed. The labeling and the labeled nuclides differ solely in their energy content.

Nuclear-analytical indicator methods belong for the most part to the type of isotope indication. The radioactive indicator is an isotope of the studied element and is in the same chemical form as the particular element. The isotopic indication refers to the use of stable and radioactive isotopes for the function of the indicator. The chemical properties of the labeling and the labeled component are very much alike, but only within the framework of the existence of isotope effects.

A special group of microindication is given by the isochemical indication in which the indicator and the element studied have similar chemical properties (alkali metals, halogens, rare earth metals, etc.).

A particular group of microindication may be formed by physical indication. It is used for the most part for labeling systems in which no chemical reaction takes place. There is a physical similarity between the indicator and the labeled substance; they need not have the same chemical form, but their reciprocal concentration ratio cannot be altered during the course of the phenomenon being followed. Physical indication includes also, the use of the universal indicator. As such we may consider radioactive isotopes of aerogens (rare gases), which although they exist in the labeled substance in solid state, are bound physically and not chemically.

Furthermore, those radioactive isotopes can be suitably used for analytical purposes that represent the part of the *reagent* which enters along with the test element or test component of the analyzed system into the common molecule.

Radioactive indicators are rarely present in elementary form; generally they are built into a molecule. In the production of radionuclides their chemical form is usually very simple and for the most part such that it is suitable from the point of view of activation nuclear reactions, ensuring also a light chemical adjustment for further utilization. In the case of inorganic systems, the radioactive isotope becomes incorporated into the molecule by way of simple, generally exchange chemical reactions.

It is a somewhat more complicated case when a labeled organic compound should be prepared. The complexity of this task resides in the variability of organic compounds, but also in the limited choice of suitable radioindicators. There are many different ways of incorporating radioactive isotopes into the molecules of organic compounds. The most significant among them are methods involving chemical syntheses, isotopic exchange reactions, activated isotopic exchange reactions, and biological syntheses.

Organic compounds are most frequently labeled by chemical syntheses. This procedure permits the radioactive atom to be incorporated into the required position of the labeled molecule. Thereby a substantial specific activity is achieved. A drawback is the time-consuming procedure required in the case of a multistep synthesis.

A very simple method of labeling a compound is through isotope exchange reaction. This procedure is based on the exchange of atoms of the same element in the molecules of various chemical compounds (that is, between atoms of the same element). Because of its simplicity the method is relatively rapid and the labeling is specific.

Isotopic exchange reactions gain in efficiency during labeling if the energy of radioisotope atoms exceeds the thermic value, or the molecule of the compound is energetically activated. The radioactive isotope which is to be incorporated into the labeled molecule may gain increased energy by recoil in the nuclear reaction, by acceleration of its ion form, or by its own radiation.

In some peculiar cases biological synthesis may be used for labeling. This is a simple method of radioactive labeling of complex biological substances.

Of broader significance is the special way of radioactive indication represented by *internal labeling*. In this procedure the radioactive isotopes are produced directly in the labeled system as a result of an activation reaction. Internal labeling by means of radioactive indicators represents the process of induced radioactivation. Activation indication is of extraordinary significance in analytical chemistry and is known as activation analysis. This process in the labeled system gives rise to the radioactive isotope of the element under study, regardless of its chemical form.

Organic molecules, too, can be labeled by the direct activation of some definite atoms of their element by means of a suitable nuclear reaction. This method is particularly applicable to the nuclear conversion of nitrogen in organic compounds to radioactive carbon.

The simplest mode of labeling compounds by the stable isotopes is by isotopic exchange reactions.

5.3.4 Distribution of a Radioactive Indicator in the Labeled System

The distribution of a radioactive indicator in the labeled chemical system is determined by the thermodynamic properties of the ensuing mixture system. The radioactive indicator is the microcomponent and the indicated system the macrocomponent. The resultant system may be looked upon as one that ensues from ideal solutions of both the components which, before being mixed, are in the same chemical state [28].

The solution is prepared by an ideal mixing of solutions AB and *AB in identical concentration (C). According to the assumption both solutions are ideal and therefore the following relation is valid for the entropy of the ion ratio of both the solutions:

$$S_{*A} = n(R \ln T - R \ln C + S^0_{*A}), \tag{5.22}$$

$$S_A = N(R \ln T - R \ln C + S^0_A), \tag{5.23}$$

where n is the number of grammolecules of the labeled system *AB, N the number of grammolecules of the non-labeled compound AB, S^0_{*A} and S^0_A are constants characterizing the entropy of substances *AB and AB, respectively.

When both solutions are mixed there ensues a mixture with partial molar concentrations of the component *AB:

$$C_{*AB} = \frac{n}{n+N} C = X_{*A} C \tag{5.24}$$

and of the component AB

$$C_{AB} = \frac{N}{n+N} C = X_A C. \tag{5.25}$$

After mixing, the entropies are as follows:

$$S'_{*A} = n(R \ln T - R \ln C - R \ln X_{*A} + S^0_{*A}), \tag{5.26}$$

$$S'_A = N(R \ln T - R \ln C - R \ln X_A + S^0_A). \tag{5.27}$$

The change in entropy during mixing will be:

$$\Delta S_{\text{mix}} = (S'_{*A} + S'_A) - (S_{*A} + S_A), \tag{5.28}$$

hence

$$\Delta S_{\text{mix}} = -n\, R \ln X_{*A} - NR \ln X_A \; \ldots \tag{5.29}$$

Entropy of mixing for gram-mole of the mixture is obtained by dividing the above values by the number of gram-moles of mixture:

$$_M\Delta S_{\text{mix}} = -X_{*A}\, R \ln X_{*A} - X_A R \ln X_A \; \ldots \tag{5.30}$$

Generalizing the above consideration we have: In mixing an arbitrary number of components, the entropy of mixing related to 1 gram-mole of mixture is given by

$$_M\Delta S_{\text{mix}} = -R \sum X_i \ln X_i \; \ldots .$$

This relation refers to any stage whatever of the mixing. Completion of mixing, hence, the state of equilibrium, characterizes the entropy constancy:

$$\delta(_M\Delta S_{\text{mix}}) = 0 \; \ldots . \tag{5.31}$$

These considerations may serve as **a** basis for determining the distribution of a radionuclide in the indicated system. In a state of equilibrium the radioactive indicator becomes uniformly distributed among the atoms of the macrocomponent in the same chemical state. An atomic fraction of a radioactive indicator in any unit volume of the mixture has the same value as is its atomic fraction for the entire indicated system.

A uniform distribution of a radioactive indicator in equilibrium within the framework of one chemical form of a macrocomponent may be spoken of even if the macrocomponent is present in the indicated system in different chemical bonds. Even if the uniform distribution of a radioactive isotope within the limits of one type of chemical bond is proved, we cannot assert—at least not in every case—that the value of uniformity in distribution is identical for all the binding types of the macrocomponent. This means that the distribution value of a radioisotope for the individual stages, or the "compartments" of the system, need not be the same.

The degree of distribution (X_i) of a radionuclide in the system may be expressed by the ratio

$$X_i = \frac{\Delta n_i}{\Delta n + \Delta N}, \tag{5.32}$$

where Δn is the number of radioactively indicated atoms or molecules of the component i in an arbitrary volume unit of the labeled system, and ΔN the number of non-labeled atoms or molecules in it.

If no complications arise in the process of mixing during the preparation of the labeled system that would be conditioned by the occurrence of several types of bonds of isotopic atoms with the radionuclide involved, then subsequent chemical conversions within the system will affect distribution of the radioactive indicator only if the chemical transformation is accompanied by isotope separation. The degree of an eventual isotope separation in current chemical processes is generally negligibly small, and therefore, isotopic mixtures prepared by an ideal mixing represent an invariant system with regard to chemical processes. Hence, the distribution value of the ith radioindicator (X_i) is constant for the ensuing phases or compartments of the labeled system.

Decisive for the use of a radioactive isotope in the function of an indicator of a chemical system is how the state of dispersion of the radioactive indicator delineates the chemical transformations or changes in the state of macrocomponent equilibrium of the labeled system. Changes of state and equilibrium properties are conditioned by the number of components in the system, by thermodynamic parameters and various further factors. If it be assumed that the overall state will be influenced by changes in only one factor while the others will be kept constant, then a distribution of the radioactive indicator will characterize changes in values of that single factor. Such a condition occurs when the distribution of the radioactive indicator corresponds to that numeric value of the entropy of mixing which represents the absolutely accessible maximum value.

Hence, isoentropic systems appear to be most suitable for radioindicator studies when quantities of substances, thermodynamic parameters and other factors determining the macroscopic state of the system are to be estimated.

Concepts on radioactive indicator distribution in the labeled system are of significance also for analytical studies. The distribution of the radioactive microcomponent in the labeled system and its changes during analytical processes may affect the measured values of sample activities. Consequently, a correct analytical interpretation of experimental results is conditioned by a knowledge of these phenomena.

5.4 Selection of an Isotope Indicator

The use of isotope indicators is based on the fact that all the elements are a mixture of isotopes. The total number of nuclides known today is close to 1500, of which more than 80 % are radioactive nuclides. Such an extensive reserve of nuclides provides an abundance of possible isotope indicators. The choice of one indicator is ruled by a whole series of criteria, more extensive for radioactive than for stable isotopes, by their nuclear radiation properties.

The determining radiation properties in the selection of radioactive isotopes as indicators – as also for a general use of radionuclides – are the type of emitted radiation, its energy and its half-life. The first two of these radiation parameters principally affect radiation detection, and the time parameter must be taken into account from the point of view of the duration of the experiment. The importance of radiation parameters in the choice of radionuclides is dealt with in detail in Chapter 4.6.

The choice between a stable and a radioactive isotope indicator is conditioned by the properties of the particular isotope. Thus, for example, many elements, very significant from an analytical point of view, have no suitable radioactive isotopes—helium, nitrogen, oxygen, etc. An isotopic indication of these elements is carried out nearly exclusively with the aid of their stable isotopes, although such a labeling is accompanied by many technical difficulties, among which should be mentioned the complex technique of obtaining stable nuclides in suitable purity, their low sensitivity and precision, and exacting procedures in the measuring technique.

A wider possibility of choice of an isotope indicator is offered by certain elements, the more significant of which are hydrogen, carbon, sulfur and some others. There exists a whole series of elements that possess but a single stable isotope (monoisotopic) or that do not eventually dispose of suitable stable isotope indicators, such as, beryllium, fluorine, sodium, aluminum, phosphorus, manganese, cobalt, arsenic, iodine, cesium, gold, etc. The use of stable isotope indicators applies predominantly to light elements and their compounds and therefore the isotope effects cannot be neglected.

Nevertheless stable isotope indicators are used—and this even when relevant element has a suitable radioactive isotope—in some specific cases:

- in the study of isotope effects in light elements and their compounds, where radioactive isotope detection would not be as precise as mass spectrometry with stable isotopes,

- in the function of an indicator, stable isotopes may be used in even higher concentrations than radioactive isotopes, as there is no danger of radiation effects, and they may be used also to indicate all the molecules in the system under study.

However, a whole series of priorities speak in favor of radioactive isotope indicators. Until the thirties of the present century their application was markedly limited. It could make use of only a few natural radionuclides of some heavy elements—suitable for chemical indication. Today, following the discovery of artificial radioactivity, the progress of nuclear sciences and nuclear technique, it is mainly due to nuclear reactors that the majority of elements possess a suitable radioisotope for the function of isotope indicator. Artificial radionuclides have had a decisive influence on the range of choice of radioactive isotope indicator, and conditions have thereby been created for the utilization of all those priorities that characterize radioactive isotope indication.

The decisive priority of radioactive indication resides in its extraordinary sensitivity of detection. The emittance of radioactive radiation and the existing technique for its measurement make it possible to count even a few radioactive atoms. That means that a small number of labeled atoms with a negligible mass can serve to indicate macroscopic systems.

When a substance of molecular weight M is isotope-indicated with a radionuclide whose half-life is T, then the mass of a radioactive carrier-free indicator ($^*m$) is given by the relation [11]:

$$^*m = {}^*N \frac{M}{N} = \frac{AT}{\ln 2} \frac{M}{N}, \tag{5.33}$$

where $^*N$ is the number of radioactive molecules, N Avogadro's number, and A the absolute radioactivity of the indicator expressed by the number of decays or—in the next relation—in Curie units (A_{Ci}):

$$^*m = 2.40 \times 10^{-24}\, MTA = 8.88 \times 10^{-14}\, MTA_{\text{Ci}}. \tag{5.34}$$

A review of the relationship between the mass of pure radioactive substance and half-life is shown in table 5.4. The data relate to the molecular weight of the labeled substance, equal to 100. Its activity, determined by 200 decays, is easily measurable by current equipment for measuring radioactivity [11].

Table 5.4

Radioindicator Mass in Dependence on Half-Life

Half-Life	Radioindicator mass (μg)
1 hr (3.6×10^3 sec)	1.73×10^{-12}
1 day (8.64×10^4 sec)	4.14×10^{-11}
1 month (2.59×10^6 sec)	1.24×10^{-9}
1 year (3.15×10^7 sec)	1.51×10^{-8}

With the aid of high-sensitive radiation detectors the degree of indication may still be reduced by about an order of 2. This extreme sensitivity of the radioindicator method makes it possible to apply it in the analysis of trace amounts of a substance.

The use of radioactive indication is further supported by the advantage of an unusual sensitivity of radioactive radiation detection. This extraordinary radiation detection sensitivity is connected with the simplicity and speed of the measuring procedures. The range of instruments available at the present time assures a rich choice of the most suitable radiation detector to achieve a high degree of detection efficiency. In the detecting and measuring techniques radioisotope indicators exceed by far the use of stable isotope indicators whose detection—except for deuterium—involves a complex and technically exacting procedure by means of mass spectrographs. Even the preparation of stable isotopes and of samples is more complicated in the case of stable isotope indicators and the measurement is time-consuming.

The analytical use of radioactive indication is less exacting as regards the chemical purity of the compound to be determined and sometimes the indicator need not even be separated from the analyzed sample. Noteworthy also is the specificity of radioactive activation from the point of view of its analytical application.

The extreme sensitivity of nuclear radiation detection permits the amount of the radioactive indicator to be reduced to a minimum. Stable isotopes, however, are employed in amounts that even in conditions of indication surpass by far those of radioactive indicators.

In some specific cases use has been made of a double microindication: indication simultaneously in one system for its two elements. The combination of the indicators refers mainly to a pair of stable and radioactive nuclides. Double indication has been reported so far in biochemistry and biology, as well as in analytical chemistry.

References

1. Sheppard G. W., *Basic Principles of the Tracer Method.* Wiley, New York 1962.
2. Hevesy G., Physikal Z. **16,** 52 (1915).
3. Hevesy G., Paneth F., Z. anorg. Chem. **82,** 323 (1913).
4. Hevesy G., *Radioactive Indicators.* Interscience, New York 1948.
5. Friedlander G., Kennedy J. W., Millar J. M., *Nuclear and Radiochemistry* (ed. 2). Wiley, New York 1964.
6. Zaborenko K. B., Iofa B. Z., Lukyanov V. B., Bogatyrev I. O., *Metod radioaktivnykh indikatorov v khimii.* Izd. Vyshshaya shkola, Moscow 1964.
7. Whitehouse W. J., Putman J. L., *Radioaktiv izotopok.* Akadémiai Kiadó, Budapest 1955.
8. Mikheyeva L. M., Mikheyev N. B., *Radioaktivniye izotopy v analiticheskoy khimii.* Gosatomizdat, Moscow 1961.
9. Etter L. E., *The Science of Ionizing Radiation, Modes of Application.* Charles C. Thomas, Springfield, Ill. 1965.
10. Haissinsky M., *La chimie nucléaire et ses applications* (Transl. into Hungarian: *A magkémia és alkalmazásai*). Akadémia Kiadó, Budapest 1963.
11. Lerch P., Chimia **21,** 1 (1967).
12. Brodský A. E., *Isotopenchemie* (German transl. from Russian: *Khimiya izotopov*). Akademie Verlag, Berlin 1961.
13. Bigeleisen J., Ann. Rev. Nucl. Sci. **2,** 221 (1953).
14. Yankwich P. E., Ann. Rev. Nucl. Sci. **3,** 235 (1953).
15. Bigeleisen J., J. Chem. Phys. **17,** 675 (1949); J. Phys. Chem. **56,** 823 (1952).

16. Wiberg K. B., Chem. Rev. **55,** 713 (1955).
17. Bernstein R. B. J., J. Phys. Chem. **56,** 893 (1952).
18. Shiner V. J. J., Jr., Am. Chem. Soc. **75,** 2925 (1953).
19. Melander L., *Isotope Effects on Reaction Rates.* Ronald, New York 1963. (Transl. into Russian, Izdatelstvo Mir, Moscow 1964.)
20. Hahn O., *Applied Radiochemistry.* Cornell Univ. Press, Ithaca 1936.
21. Paneth F., *Radioelements as Indicators.* McGraw-Hill, New York 1928.
22. Siri W. E., *Isotopic Tracers and Nuclear Radiations.* McGraw-Hill, New York 1949.
23. Majer V., *Základy jaderné chemie (Principles of Nuclear Chemistry).* SNTL, Prague 1961.
24. Rotschild S., *Advances in Tracer Methodology.* Plenum Press, New York 1965.
25. Kamen M. D., *Isotopic Tracers in Biology.* Academic Press, New York 1957.
26. Rescigno A., Biochem. Biophys. Acta **15,** 340 (1954).
27. Rescigno A., Biochem. Biophys. Acta **21,** 111 (1956).
28. Imre L., *Rádioaktív nyomjelzös módszerek elvi alapjai.* Országos atomenergia bizottság, Budapest 1961.
29. Gorsuch T. T., *Radioactive Tracers in Chemical Analysis.* The Radiochemical Centre, Amersham 1966.
30. Lambie D. A., *Techniques for the Use of Radioisotopes in Analysis.* E. and F. N. Spon, Ltd., London 1964.

List of Abbreviations and Symbols

A	Mass (nucleonic) number; nuclear mass; activity; number of disintegrations; constant
B	Constant
Ci	Curie
D	Enrichment (crystallization) factor; dose
D_{rad}	Physical dose
D_{rem}	Biologically effective dose
E	Energy; intensity of the electric field
E_{max}	Maximum energy of beta particles of a given species emitted by a given radionuclide
EC	Electron capture
I	Isotopic number; intensity of radiation; relative activity; counting rates; ionization potential
K	Equilibrium constant
K_a	Dissociation constant
K_D	Distribution coefficient
K_s	Stability constant
L	Solubility
M	Molecular (atomic) weight
M	Molar; gram mole/litre
MeV	Million electron volts
N	Neutron number; Avogadro's number; No. of atoms; No. of gram molecules
N_0	No. of atoms present at zero time
N	Normality
P	Reaction product
Q	Energy of the nuclear reaction (Q value)
R	Range; square weight of the absorber; absorber thickness; resistance
$R_{\frac{1}{2}}$	Half-thickness
RBE	Relative biological effectiveness
S	Chemical system
S	Specific activity
T	Half-life; temperature

T_{Ef} Effective half-life
T_B Biological half-life
U Voltage
V Volume
X_i Degree of distribution of a radionuclide in the system
Z Atomic number
a Counting rate related to the unit of channel width
b Barn
c Velocity of light; concentration
cpm Counts per minute (counting rate)
cps Counts per second
d Deuteron; thickness; distance; range
$d_{\frac{1}{2}}$ Half-thickness
dpm Disintegration per minute
e Electron
f Partition function
g Gram
h Planck's constant
hr Hour
k Boltzmann's constant; constant; rate constant for a reaction
l Distance; litre
m Mass
m Meter
m. u. Mass unit
mCi Millicurie
mg Milligram
mb Millibarn
min Minute
n Neutron; No. of particles; No. of channels
n_p No. of ion pairs
nCi Nanocurie
p Proton
r Radius; distance
r Roentgen
$r_{\frac{1}{2}}$ Half-width value
s Constant
sec Second
t Time
v Velocity
x No. of atoms; No. of events in a given time
y No. of atoms
yr Year
α Alpha particle; ratio of the rate constants
β Beta particle; ratio of the relative velocity of the beta particles and the speed of light ($\beta = v/c$)
γ Photon
ε_n Neutron flux of a given energy, which causes the maximally permissible dose intensity
η Proportionality factor
λ Proportionality constant (decay constant); wavelength of neutron or gamma ray; logarithmic crystallization constant

μ	Mass absorption coefficient
μ'	Linear absorption coefficient
μb	Microbarn
μg	Microgram
μCi	Microcurie
μsec	Microsecond
ν	Neutrino; frequency of electromagnetic radiation
ϱ	Density
σ	Nuclear (microscopic) cross-section; linear absorption coefficient for Compton scattering
$\varkappa$	Linear absorption coefficient for pair production
τ	Average life; lifetime; dead time; linear absorption coefficient for photoelectric effect
Φ	Neutron flux
Γ	Level width
Σ	Macroscopic cross-section
Δ	Deviation; difference

Name Index

Subject Index